AF711549

THE MEDIEVAL LATIN TRANSLATION OF THE *DATA* OF EUCLID

The Medieval Latin Translation of the Data of Euclid

Shuntaro Ito

Springer Japan KK

The publication of this book was supported by a Grant-in-Aid for Publication of Scientific Research Results from the Ministry of Education, Science and Culture, Japan.

Library of Congress Cataloging in Publication Data

Euclides.
The Medieval Latin translation of the Data of Euclid.

Bibliography: p.
1. Geometry – Early works to 1800. I. Itō, Shuntarō, 1930– II. Title.
QA31.E8613 1980 516.2 80-14257
ISBN 978-1-4899-4967-7 ISBN 978-1-4899-4965-3 (eBook)
DOI 10.1007/978-1-4899-4965-3
CIP-Kurztitelaufnahme der Deutschen Bibliothek

Euclides:
[The medieval Latin translation of the Data]
The medieval Latin translation of the Data of Euclid / Shuntaro Ito. – Boston, Basel, Stuttgart: Birkhäuser; [Tokyo]: University of Tokyo Press, 1980.
Einheitssacht.: Data <engl.>
ISBN 978-1-4899-4967-7
NE: Ito, Shuntaro [Hrsg.]

© Springer Japan 1980
Originally published by University of Tokyo Press in 1980
Softcover reprint of the hardcover 1st edition 1980
UTP 3041-66731-5149

All rights reserved. No part of this publication may be reproduced or transmitted in any form or by any means, electronic or mechanical, including photocopy, recording, or any information storage and retrieval system, without permission in writing from the publisher.

FOREWORD

It is a well-known truth that modern mathematics takes its departure from Greek mathematics. But less well known are the details of the transmission of the texts of Greek mathematics to the West. Particularly complex was the transmission of the *Elements* of Euclid, and, as Professor Ito points out in his preface, several scholars have made important beginnings in that direction. To these beginnings we must now add this text of the medieval Latin version of the *Data* of Euclid edited by Professor Ito. Ito's text is based on all four of the extant manuscripts, though it leans most heavily on the Oxford MS. It is of considerable interest that this medieval Latin version was made from a Greek text that was, in the main, from the Pre-Theonine tradition, though occasionally the Latin version reveals agreement with the Theonine text. I believe Professor Ito is correct in his assumption that the text is uncontaminated by the purported translation of the *Data* from Arabic made by Gerard of Cremona since there are no evident Arabicisms in it and it agrees so closely in word order and grammatical form with the Greek text. But we cannot know this with absolute certainty until some copy of Gerard of Cremona's translation turns up and a careful comparison is made.

The Latin *Data* appears to have been made by the translator who produced the extant medieval translations from the Greek of Euclid's *Optics* and *Captoptrics* and Proclus' *Elementatio physica*, since all of these translations display the same translating techniques. However, it is very much an open question as to whether the translator of these four works is identical with the anonymous translator from the Greek of Ptolemy's *Almagest* and Euclid's *Elements*. Though Ito seems to feel that the translators might be the same, he admits the cogency of John Murdoch's arguments that would distinguish the two men. Though the question may never be settled, a fuller examination of the texts and of the *Elements* and the *Almagest* and a detailed comparison of the techniques used in translating technical terms (and not merely particles, conjunctions and the like) in these longer works with those found in the *Data* may bring us closer to the solution. In this connection may I say that the more I have studied medieval translating techniques the less confident I am that the study of com-

parative translations of particles and conjunctions gives us anything more than a start toward the identification of the translator. If the identity of the two translators could be established, then, as Ito suggests, it would be a fair conclusion that the translation of the *Data* was done in Sicily prior to 1160. Finally, the reader should note that Ito briefly studied the Renaissance translations of the *Data* made from the Greek by Giorgio Valla and Bartolomeo Zamberti, the first a loose and incomplete version, the second a complete translation of the Theonine version. Ito distinguishes these translations from the medieval text he has edited.

It is evident, then, that the student of medieval mathematics should find most helpful the critically edited version of the *Data* which has been presented by Professor Ito.

Marshall Clagett
The Institute for
Advanced Study
Princeton, N. J., U.S.A.

PREFACE

It is a pressing desideratum in the history of medieval science to have the whole corpus of *Euclides Latinus* in a modern edition. There is a sense in which one can say that this task constitutes a cornerstone of the study of scientific transmission in the Middle Ages.

While the efforts of such eminent historians of mathematics as Cantor, Zeuthen, Tannery, Heiberg, Hultsch and Heath have made amply clear the *Greek* tradition of Euclid, and the studies of experts like Wenrich, Wüstenfeld, Klamroth, Steinschneider, and Suter have shed some light on the *Arabic* tradition of Euclid's works, much confusion still remains concerning the *medieval Latin* tradition of Euclid. Recently Marshall Clagett has dispelled some of the darkness and provided a right start on the problem of the "medieval Euclid" by investigating the newly collected manuscripts of medieval versions of Euclid.

Under his direction, various medieval Latin versions of the *Elements* translated from Arabic were examined and edited by his former students George Goldat and St. Martin van Ryzin. H. L. L. Busard also published the Latin text of the *Elements* translated from the Arabic by Hermann of Carinthia(?), and M. Folkerts issued the Boethian texts of the *Elements*. John Murdoch discovered manuscripts of the Latin version of the *Elements* from the Greek and did an excellent study of it.

As regards the medieval versions of *other* works of Euclid, however, nothing as yet has been done besides the editing of the Latin *Optics* by J. L. Heiberg and more critically by W. R. Theisen.

In the present work I have edited the medieval Latin version of the *Data* from photographs of the four extant manuscripts: those at Oxford, Paris, Dresden, and Berlin. Inasmuch as this is the first edition of the Latin version of the *Data* of Euclid, I hope that it will be a step toward the completion of a corpus of *Euclides Latinus*.

Further, I have translated this edited Latin text into English. This will be the first modern English version of the *Data*. As is well known, Simon's famous paraphrase of the *Data* not only was based on a vitiated Greek text, but also changed the original content so much by his own interpretation that it could hardly be called a proper translation.

To the collated Latin text of the *Data* and its English translation, I have also added preliminary studies on the significance and the history of this work and particularly on the preparation of the medieval Latin translation of the *Data*. Although I am conscious of the fact that this part of my study, especially the problem of identification of the translator, still leaves some problems, I believe that by this investigation I have been able to shed light on some aspects of considerable importance for the history of scientific transmission in the twelfth century.

I started my study of this medieval version of Euclid's *Data* in 1961, when I was Research Assistant to Professor Clagett at the Institute for Research in Humanities, the University of Wisconsin. At the time I presented the result of my study in 1963 as a Ph.D. dissertation, only two manuscripts (Oxford and Paris) were available for collation of the text. Afterwards I emended the text at the Institute for Advanced Study, Princeton, in 1971 and 1972, taking into consideration two more extant manuscripts (Dresden and Berlin) of which I could avail myself at Princeton, and completed the final text of this Latin version of the *Data*. At the institute I also revised my investigation of the preparation of this translation and made further study of the Latin tradition of the *Data*, especially during the Renaissance; the results are included here. Therefore this work is not identical with my Wisconsin dissertation: many emendations and improvements have been made both in the text and in the historical study.

Now that I am to publish this work, my deepest thanks are due, first of all, to Professor Marshall Clagett of the Institute for Advanced Study, who as my master in the history of science at Wisconsin gave me constant generous assistance and helpful direction and also later enabled me to continue my study at Princeton. I would not have been able to complete this work without his invariable aid and kind encouragement. I would also like to express my appreciation to Mr. Tohru Sato for his preparation of the index for this book. Appreciation is also due Ms. Etsuko Hamao of the University of Tokyo Press, who took care of the book's publication.

Finally, I would like to offer a word of thanks to my wife for her inexhaustible help in our life at Madison and Princeton, where I made efforts toward completing this work.

Shuntaro Ito

October 1979

CONTENTS

THE MEDIEVAL LATIN TRANSLATION OF THE *DATA* OF EUCLID

INTRODUCTION

Basic Aim and Structure of This Work

The basic aim of this work is to present the medieval Latin version of the *Data* of Euclid and its English translation. To edit this Latin version of the *Data*, I used the following four manuscripts:

1) MS. Oxford, Bodl. Auct. F.5.28, 13c, 99r-113r.
2) MS. Paris, Bibl. Nat. lat. 16648, 13c-14c, 60r-91v.
3) MS. Dresden, Sächs. Landesbibl. Db 86, 14c. 200r-213r.
4) MS. Berlin, Staatsbibl. Preuß. Kulturbesitz Q510, 14c, 178v-191r.

These four are the only extant manuscripts of the Latin translation from the "Greek" of the *Data*. It seems that no other manuscripts of this Latin version of the *Data* are found. Therefore the text I presented here is edited with collation of all manuscripts now available.

My English version of the *Data* is, I believe, much closer to Euclid's original text than previous English versions of it, because the Latin translation in these manuscripts seems to have come from an older and better Greek manuscript than Theon's recension. Actually this Latin version accords very well with Menge's Greek text which was edited on the basis of the non-Theonine manuscripts, chiefly MS Vat. 190 discovered by Peyrard in 1808.

It seems, however, necessary to introduce the text with a historical analysis of the *Data*, for not enough is known about the work itself or about the historical transmission of it. In fact, study of the *Data* has been very much neglected, compared with that of the *Elements*. Hence, in Part I I have reorganized the scattered studies concerning the *Data* and also made clear the historical transmission of the work. In particular, in my study of the preparation of the medieval Latin translation of the *Data*, I have added some new facts about scientific transmission in twelfth-century Sicily.

Part I consists of three chapters:

Chapter 1 delineates the place of the *Data* in Euclid's work and the significance of this work for Greek mathematics.

Chapter 2 will follow the history of the *Data* in its three traditions—Greek, Arabic, and Latin. Here are mentioned the modifications, recensions, redactions, and translations—Arabic and Latin—and also the mo-

dern editions of this work. Information about the extant manuscripts of the *Data* is also given.

Chapter 3 treats the problem of the preparation of the medieval Latin translation from Greek of the *Data*. This problem is divided into three sub-problems:

1) When and where did this Latin translation take place?
2) Who translated it into Latin?
3) How were the four extant manuscripts derived from the original and how are they different from one another?

Each of these problems, and especially problem (2), is difficult to solve completely. But I can say at this stage of study that I have formed tentative conclusions not unimportant for the history of the transmission of scientific works in the early Middle Ages. In order to test the conclusions more fully, some other manuscripts ought to be consulted. But this is a part of a large study of Sicilian mathematical translations that goes beyond the limits of this study.

Following a comparative description of the four extant manuscripts, I have inserted specimen plates for the reader's reference.

Part II also consists of three chapters:

Chapter 4 contains general remarks on the manner in which I have edited the text and translated it into English.

Chapter 5, the main portion of this book, gives the collated Latin text of the *Data* and its English translation on parallel pages.

Chapter 6 contains a commentary on the text, explicating particular and peculiar notions in the text of the *Data*.

Last, I have included a comprehensive bibliography.

PART 1

HISTORICAL EXPOSITION OF THE *DATA* OF EUCLID

CHAPTER 1

EUCLID AND THE *DATA*

As is the case for other Greek mathematicians, we have only the most meager particulars of the life of Euclid. Almost all that is known about him is contained in a famous short passage[1] of the *Commentary* of Proclus. From Proclus' remark, we know only that Euclid came after the pupils of Plato and before Archimedes; therefore, he must have flourished c. 300 B.C. when the first Ptolemy reigned in Egypt.

Although little is known of his life, his achievement in mathematics is, of course, well known from his monumental *Elements*, which gave us for the first time in history a completely rigorous deductive system of pure geometry. As an incarnation of mathematical reasoning, it has been greatly admired and very much studied. In fact, no other mathematical work has been so much studied and so often commented upon as the *Elements*.

The *Data*, which has been much less well studied, is among Euclid's other genuine works the one most closely connected with the *Elements*, since it too deals with plane geometry, the subject matter of Books I-VI of the *Elements*. That this is a genuine work of Euclid is commonly accepted because of its close relation to the *Elements* and its use of Euclidian style. In fact, its authenticity was attested a long time ago by Pappus,[2] who regarded this work as important enough to be included in the "Treasury of Analysis."

The object of the *Data* is to provide a series of propositions such as to prove that, if in a given figure certain parts or relations are given, other

1) *Procli Diadochi in primum Euclidis elementorum librum commentarii*. ed. G. Friedlein, Leipzig (1873), p. 68, 6–20.

οὐ πόλυ δὲ τούτων νεώτερός ἐστιν Εὐκλείδης ὁ τὰ στοιχεῖα συναγαγὼν καὶ πολλὰ μὲν τῶν Εὐδόξου συντάξας, πολλὰ δὲ τῶν Θεαιτήτου τελεωσάμενος, ... γέγονε δὲ οὗτος ὁ ἀνὴρ ἐπὶ τοῦ πρώτου Πτολεμαίου· καὶ γὰρ ὁ Ἀρχιμήδης ἐπιλαβὼν καὶ τῷ πρώτῳ μνημονεύει τοῦ Εὐκλείδου, ... νεώτερος μὲν οὖν ἐστι τῶν περὶ Πλάτωνα, πρεσβύτερος δὲ Ἐρατοσθένους καὶ Ἀρχιμήδους. οὗτοι γὰρ σύγχρονοι ἀλλήλοις, ὥς πού φησιν Ἐρατοσθένης".

2) Cf. *Pappi Alexandrini collectionis quae supersunt* volumen II, ed. Fr. Hultsch, Berlin (1877), pp. 636–640.

parts or relations are also given in one or another of the senses initially defined.[3)]

So this work begins with definitions of the senses in which things are said to be "given" (*data*). Things such as spaces, lines, and angles are said to be "given in magnitude," when we can assign equals to them (def. 1). On the other hand, rectilineal figures are said to be "given in species," when each of their angles is given and also the ratios of their sides one to another are given (def. 3). Further, points, lines, and angles are said to be "given in position" when they always hold the same place (def. 4). Furthermore, a circle is said to be "given in position and magnitude" when its center is given in position and its radius is given in magnitude (def. 6).

Following these definitions, we find propositions which deal with magnitude or quantity (prop. 1–23), lines (prop. 24–38), rectilineal figures (prop. 39–86), and circles (prop. 87–94) in turn. To show their character, I give here only a few examples:

Proposition 4: If a given magnitude is taken from a given magnitude, the remainder is also given (in magnitude).

Proposition 25: If two lines given in position intersect, the point at which they intersect each other is also given in position.

Proposition 39: If each side of a triangle is given in magnitude, the triangle is given in species.

Proposition 91: If outside a circle given in position there is taken a given point and from that point there is drawn a certain straight line into the circle, the rectangle comprised by the straight line so drawn and the straight line which is between the point and the convex circumference is given.

In these propositions, the word "given" is employed in two senses. It means first "actually given" and second "given by implication." Propositions are all to this effect, that a certain partial description of a certain magnitude or of a certain geometrical figure implies a more complete description, just as the description of a triangle as equilateral implies its description as equiangular. We can say that propositions in the *Data* bring what is given by implication to what is given actually. In the course of these proofs, the *Data* uses propositions of the *Elements* as well as

3) Cf. Thomas L. Heath, *A History of Greek Mathematics*, Vol. I, Oxford (1921), p. 422.

propositions already proved in the *Data*, as is shown *e.g.*, by the proof of Proposition 91 referred to above.[4]

Thus it is clear that the *Data* presupposes the *Elements*. The *Data* is, in fact, nothing but "a series of easy riders on the *Elements*."[5] If this is so, why did Euclid write this particular work? What function or significance does this work have for Greek mathematics?

The answer to these questions lies in the special role which the *Data* plays in the process of "analysis," one of the most important methods of Greek mathematics.

"Analysis" was the one general method which the Greeks used for solving "the more abstruse problems" (*τὰ ἀσαφέστερα τῶν προβλημάτων*).[6] As Heath says, "We have, let us suppose, to construct a figure satisfying a certain set of conditions. If we are to proceed at all methodically and not by mere guesswork, it is first necessary to 'analyse' those conditions. To enable this to be done we must get them clearly in our minds, which is only possible by assuming all the conditions to be actually fulfilled, in other words, by supposing the problem solved. Then we have to transform those conditions into other conditions which are necessarily fulfilled if the original conditions are, and to continue this transformation until we at length arrive at conditions which we are in a position to satisfy. In other words, we must arrive at some relation which enables us to *construct* a particular part of the figure which, it is true, has been hypothetically assumed and even drawn, but which nevertheless really requires to be *found* in order that the problem may be solved."[7]

Actually, it is from this moment that the particular part of the figure is "given" in the domain of the *Data*, and another part of the figure can be found to be "given" by using the propositions of the *Data*. Through this chain of steps, we get at last to the data or datum which gives a possible beginning of the process of "synthesis." Hence the end of the first part of

4) The proof of Proposition 91 is this:
From the given point *D*, let the straight line *AD* touch the given circle *ABG*. Then *AD* is given in position and magnitude.(*Data* 90) Since *AD* is given, AD^2 is also given. And $AD^2 = BD \times DG$. (*Elements* III, 36) Therefore $BD \times DG$ is given.

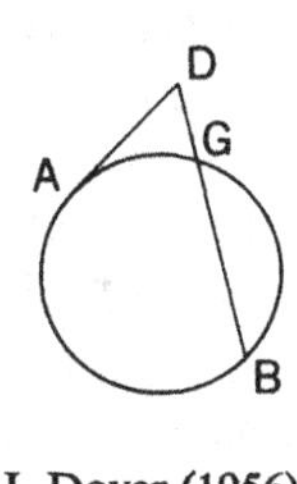

5) James Gow, *A Short History of Greek Mathematics*, New York (1923), p. 210.

6) Proclus, *Commentarii*, p. 242, 16–17.

7) Thomas L. Heath, *The Thirteen Books of Euclid's Elements*, Vol. I, Dover (1956), Introduction, Chapter IX, 6, pp. 140–1.

"analysis," the part which Hankel called "transformation,"[8] constitutes the starting point of the problem of the *Data*. In the following part of the analysis, which Hankel called "resolution," the propositions of the *Data* play their role in forging the chain of data until we arrive at the point which enables us to begin the process of "synthesis."

What has been said above will be well illustrated by the example of proposition 105 in Chapter 7 of Pappus' *Mathematical Collection*.[9] The problem is this: "Given the position of a circle *ABG* and of two points *D*, *E* outside it, it is required to draw from *D*, *E* the straight lines *DB*, *EB* cutting the circle in *B* and produced to *G*, *A* so that *AG* shall be parallel to *DE*."

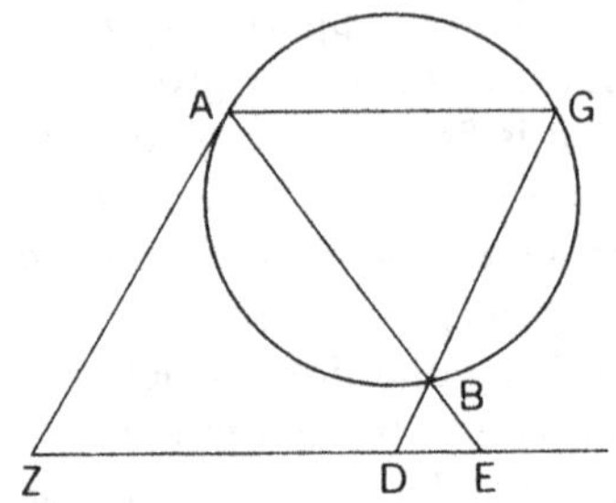

After supposing that this problem is solved, the analysis proceeds as follows: "Let the figure be drawn and also the tangent *ZA*.

[TRANSFORMATION] Then, since *AG* is parallel to *DE*, the angle at *G* is equal to the angle *GDE*. It is also equal to the angle *ZAE* (*Elements* III, 32). Therefore the angle *ZAE* is equal to the angle *GDE*, and the points *A*, *B*, *D*, and *Z* lie on the circumference of a circle, and the rectangle *AE*, *EB* is equal to the rectangle *ZE*, *ED*. But $AE \times EB$ is given, because it is equal to the square of the tangent, drawn to the circle from *E* (*Elements* III, 36).

[RESOLUTION] Therefore $ZE \times ED$ also is given, and since *ED* is given, *ZE* also is given in magnitude (*Data*, 57). But *ZE* is also given in position, so *Z* is also given in position (*Data*, 27). Now *ZA* is the tangent from a given point *Z* to a circle *ABG* given in position. Therefore *ZA* is given in position and magnitude (*Data*, 90). And *Z* is given. Therefore *A*

8) Hermann Hankel, *Zur Geschichte der Mathematik in Altertum und Mittelalter*, Leipzig (1874), p. 141. Also cf. Heath, *Op. cit.*, *ibid.*

9) Pappus, *Collectio quae supersunt*, Vol. II, pp. 830–2.

For a French translation, see Paul Ver Eecke, *Pappus d'Alexandrie, La Collection Mathématique*, Vol. II, Paris (1933) pp. 640–2. Also cf. Heath, *Op. cit.*, pp. 141–2.

is given, (*Data*, 27). But E is also given. Therefore the straight line AE is given in position and magnitude (*Data*, 26). And the circle ABG is given in position. Therefore the point B is also given in position (*Data*, 25). But the points D, E are also given in position. Therefore the straight lines DB, BE are also given in position."

Then follows the "synthesis": "Join E, D and produce it to Z so that the rectangle ZE, ED is equal to the square of the tangent drawn to the circle from E. From Z draw the tangent ZA and join A, E, etc. etc."

As is clear from this example, the existence of a systematic collection of the propositions of the *Data* does facilitate and shorten the process of "analysis" in its "resolution" part. Actually, provided that we know that a certain part of the figure is given, it is not necessary to carry out the actual operation in order to find that another part is also given. Propositions in the *Data* serve as the means for that. The *Data*, in fact, seems to have been widely used and circulated among Greek mathematicians to facilitate the procedure in the analysis indispensable to the systematic solution of geometrical problems.

Here then is the real function of the *Data* in Greek mathematics: it was a necessary *Hilfsmittel* to "analysis" which was, as I have said, one of the most important methods of Greek mathematics.

As a matter of fact, Pappus described the *Data* as forming part of the "Treasury of Analysis" (τόπος ἀναλυόμενος).[10] According to him, this was a name for a special department of geometry, "a particular body of doctrine prepared for those who, after finishing the ordinary *Elements*, wish to acquire the power of solving problems which may be set to them, involving the construction of lines."[11] In the same place, Pappus says that this τόπος ἀναλυόμενος was contributed to by Euclid, Apollonius of Perga, and Aristaeus the Elder. A little later, enumerating as many as 33 books of Greek mathematics which were devoted to the τόπος ἀναλυόμενος, he put at the top of them the *Data* of Euclid.[12]

10) Pappus, *Collectio*, Vol. II, p. 634.

By the way, the Greek word τόπος here does not mean simply a locus, but has its Aristotelian meaning of "store-house". Actually, the τόπος ἀστρονομούμενος at the beginning of Book VI of Pappus' *Collection* means the "Astronomical Treasury" consisting of books which he later discusses. Therefore τόπος ἀναλυόμενος should mean the "Treasury of Analysis" and the translation of this word as "locus resolutus" (Commandino) or as "lieu résolu" (Ver Eecke) seems not to be right. See J. Gow, *Op. cit.*, p. 211.

11) Pappus, *Op. cit.*, p. 634.

12) Pappus, *Op. cit.*, p. 638.

Heiberg evaluated the *Data* positively as belonging to higher geometry (in der höheren Geometrie),[13] while Cantor remarked that it was only an exercise book for refreshing the *Elements* (Übungssätze zur Wiederauffrischung der Elemente).[14]

But the real significance of the *Data*, as was seen above, lies in the fact that it was a very helpful collection of propositions which greatly facilitated the process of "analysis" when he who had mastered the *Elements* undertook fresh problems.

13) Johan L. Heiberg, *Litterargeschichtliche Studien über Euklid*, Leipzig (1882), p. 39.

14) Moritz B. Cantor, *Vorlesungen über Geschichte der Mathematik*, Vol. I, 3rd ed., Leipzig (1907), p. 284.

CHAPTER 2
A SURVEY OF THE HISTORY OF THE *DATA*

In this chapter I follow the historical fate of the *Data* in its three traditions, Greek, Arabic, and Latin, and mention its modern editions and translations.

Not long after Euclid wrote the *Data*. the work suffered many additions, alterations, and rearrangements. In the first place, it is clear that Apollonius of Perga concerned himself with this work, because one of the ancient scholia on the *Data* stated expressly that the last three definitions in the opening pages of the *Data* had been added by Apollonius.[15] And the additions of other kind of proofs with the mark *ἄλλως* appeared in early times. These are probably attributable to Apollonius and other Greek mathematicians.[16]

We have already seen that in the first part of the 4th century A.D., Pappus made mention of this work in Book VII of his *Mathematical Collection* as one of the most important works in the "Treasury of Analysis." He gave fairly detailed description of the contents of the *Data*. This description shows that there was a considerable difference between the text used by him and that which is extant. Although our text corresponds to Pappus' description through Proposition 62, the succeeding propositions do not agree exactly in number and order. Pappus enumerated only 90 propositions instead of 94.[17]

In the same century, Theon of Alexandria undertook his recension of the *Data* as well as of the *Elements*. He did not attempt to recover the original form of the text, but tried only to offer a readable and easily understandable text. In doing so, he not only gave explanatory additions of his own, but he also made many other alterations and free abbreviations. As a result, the Greek text of the *Data* edited by Theon represented a transformation of the original.[18] This edition of Theon was widely used

15) *Euclidis Data* cum Marini commentaria et scholiis antiquis, *Euclidis Opera Omnia* Vol. VI, ed. H. Menge, Leipzig (1896), p. 264. The ancient scholium to Definition 13, *τούτους 'Απολλωνίου φασὶν εἶναι τοὺς τρεῖς ὅρους.*

16) These other proofs are also included in Menge's edition as an Appendix. See *Euclidis Opera Omnia* VI, p. 323 seqq.

17) Pappus, *Collectio quae supersunt*, Vol. II, pp. 638–640. (Paul Ver Eecke's French translation, pp. 478–480)

18) Cf. *Euclidis Opera Omnia*, Vol. VI, Prolegomena (by Menge) p. xxxii seq. and xl.

by scholars. In fact, it constituted the basis of the modern edition of the Greek text until Peyrard discovered in 1808 an older and more original manuscript than those containing the Theonine text.

In the fifth century, Marinus of Neapolis, a student and successor of Proclus at Athens, wrote a famous commentary or prolegomena to the *Data*,[19] contributing further to the understanding of this work.

Proclus in the same century cited the substance of the first four definitions of the *Data* in his *Commentary* on the First Book of the *Elements*. In addition to distinguishing the three Euclidian ways in which things are given—namely, in "position," in "magnitude," and in "species"—he added another way of things being given, *i.e.*, given in "ratio."[20] Eutocius also used the *στοιχείωσις τῶν δεδομένων* in his edition of Archimedes' *De sphaera et cylindro*.[21]

The first Greek printed text of the *Data* (*Εὐκλείδου δεδομένα*) was brought out by Clément Hardy at Paris in 1625.[22] Then followed the Oxford edition in 1703 by David Gregory, who collected into one volume the complete works of Euclid including the *Data*.[23] Neither of them was concerned with producing a "critical" text and neither succeeded in obtaining good Greek manuscripts. Both editions were based on the vitiated Theonine manuscripts.

The important step towards a sound Greek text was taken by F. Peyrard, who between 1814 and 1818 published the *Elements* and the *Data* in Greek, Latin, and French.[24] In 1808, when Napoleon was having valuable manuscripts selected from Italian libraries, Peyrard happened to discover among the Italian manuscripts sent to Paris a very excellent manuscript of Euclidean works—cod. Vat. Gr. 190, which contained the *Elements* I-XIII and the *Data* with Marinus' commentary and ancient

19) Marinus' commentary is also included in Menge's edition of the *Data*. See *Euclidis Opera Omnia*, VI, p. 234 seqq.

20) Proclus, *Commentarii*, p. 205, 13 seq. Cf. Heath, *The Thirteen Books of Euclid's Elements*, Vol. I, p. 133.

21) For further information about the authors who cited the *Data*, cf. Heiberg, *Litteraturgeschichtliche Studien*, pp. 221–224.

22) Cf. *Euclidis Opera Omnis*, VI, Praefatio (by Menge) p. v.

23) *Εὐκλείδου τὰ σωζόμενα, Euclidis quae supersunt omnia. Ex recensione Davidis Gregorii M. D. Astronomiae Professoris Saviliani et R. S. S. Oxoniae, e Theatro Sheldoniano*, An. Dom. MDCCIII.

24) *Euclidis quae supersunt. Les Oeuvres d'Euclide, en Grec, en Latin et en Français d'après un manuscrit très ancien, qui était resté inconnu jusqu'à nos jours.* Par F. Peyrard. 3 vols. Paris, 1814–1818.

scholia on it. It proved to be not Theonine but a more ancient and original manuscript. Adopting many of the readings in this manuscript and referring to other manuscripts, Peyrard published the first critical text of the *Data* in the third volume of his Euclid edition. Although he pointed the way towards a better text, he still adopted the fallacious procedure of simply correcting the text of Gregory by using the newly discovered manuscript rather than rejecting that text altogether and starting afresh.

It was Heiberg and Menge who, completely rejecting the Oxford edition of Gregory, undertook to edit anew the complete text of Euclid on the basis of the best manuscripts then available to them. Their monumental edition of Euclid's works, *Euclidis Opera Omnia*, comprising eight volumes plus a supplementary ninth volume, was published between 1883 and 1916.

H. Menge, who edited the *Data* in the sixth volume of this *Opera Omnia*, based his text primarily on six manuscripts,[25] and the text reflects the importance of cod. Vat. Gr. 190 discovered by Peyrard and its tradition as against the tradition of the Theonine manuscripts. Therefore Menge adopted the Vatican manuscript P as the main basis of his edition. But, at the same time, note that the text of the *Data* in non-Theonine manuscripts represented by the Vatican manuscript do not show so much divergence from Theonine manuscripts as does that of the *Elements*.[26] It is also suggested by comparison with Pappus' reference and with Arabic versions that the text in extant Greek manuscripts is not the same as the original, which seems to have been altered even before Theon.[27]

With regard to Arabic versions of the *Data*, we are informed by Steinschneider that Isḥāq ibn Ḥunayn translated the *Data* into Arabic with the title of *al-Muʿṭayāt* at the end of the 9th century and that, afterwards, Thābit ibn Qurra revised this translation and also wrote a work of his own with the same title.[28] These Arabic versions of the *Data* seem to be based on non-Theonine texts. Furthermore, extant Arabic manuscripts

25) Cf. *Euclidis Opera Omnia*, Vol. VI, Praefatio, p. v and also Prolegomena, p. xiii seq. The six manuscripts are cod. Vat. Gr. 190, 10c (= P), cod. Vat. Gr. 204, 10c (= Vat.), cod. Vat. Gr. 1038, 13c (= v), cod. bibliothecae communalis Bononiensis A 1, 18, 19 signatus, 11c (= b), cod. Florentin. Laurentianus XXVIII, 1, 14c (= a) and cod. Paris, Gr. 2448, 14c (= z).

26) Cf. J. L. Heiberg, *Litterargeschichtliche Studien über Euklid*, Leipzig, 1882, p. 224.

27) Cf. C. Thaer, *Die Data von Euklid*, Springer, 1962, p. 64.

28) Moritz Steinschneider, *Die arabischen Übersetzungen aus Griechischen*, new edition, Graz (1960), p. 171. Also cf. F. J. Carmody, *The Astronomical Works of Thabit B. Qurra*, Berkley (1956), p. 221 and C. Thaer, "Euklids Data in Arabischer Fassung," *Hermes* Vol. 77 (1942).

contain a redaction of the *Data* by aṭ-Ṭūsī with his commentary in the 13th century.[29]

It is well-known that Gerard of Cremona translated this work from Arabic into Latin in the latter part of the 12th century.[30] But his Latin translation of the *Data* from Arabic seems to be lost. In this connection I should mention the fact that a Latin reworking of the *Data* found in MS Dresden Db 86, which Hultsch presumed to have come from an Arabic version, has exactly the same content as our Latin translation of the *Data* from Greek. The tract, which he earnestly wished to publish, is actually nothing but the Latin version of the *Data* supplied in this work.[31]

As far as the medieval Latin translation of the *Data* from Greek is concerned, I will investigate this in detail in the next chapter. Here I can simply note that there are four extant manuscripts of it, namely, MS Oxford, Bodl. Auct. F.5.28, MS Paris, Bibl. Nat. lat. 16648, MS Dresden Db 86, and MS Berlin, Staatsbibl., Preuß. Kulturbesitz Q 510, though Björnbo mentioned only two manuscripts—those of Oxford and Paris.[32] These manuscripts have provided the basis for my text as presented in this work. That this Latin translation of the *Data* was from Greek rather than from Arabic is quite evident from several facts I will discuss later. These anonymous manuscripts contain the identical Latin translation which seems to have been made in the 12th century, as I conjecture also in the next chapter. This Latin version agrees very well with the Greek text edited by Menge, though sometimes it coincides with Theonine texts.

As regards Renaissance Latin translations, we have the two versions of

29) Heinrich Suter, "Die Mathematiker und Astronomer der Araber und ihre Werke", *Abhandlungen zur Geschichte der Mathematik*, Heft X (1900), p. 151. The extant MSS of the redaction of the *Data* by aṭ-Ṭūsī which have been so far discovered are the following: 1) MS Berlin, Ahlwardt 5929, 2) MS Florence, Pal. 271, 273, 286. 3) MS Oxford II, 875 & 895. 4) MS Leyden 1029 etc. 5) MS Kairo, Khedive V. 200. See other manuscripts in Carmody, *Op. cit.*, p. 217.

30) M. Steinschneider, *Op. cit.*, p. 171. Also cf. his *Die europäischen Übersetzungen aus Arabischen bis zum 17. Jahrhundert*, new edition, Graz (1956), p. 18.

31) Cf. Friedrich Hultsch, "Eukleides", in Pauly-Wissowa's *Die Real-Encyclopädie der classischen Altertumswissenschaften*, new edition, Vol. VI, Pt. I, Stuttgart (1907), p. 1043. "Auch findet sich eine vermutlich aus dem Arabischen geflossene lateinische Bearbeitung der Data im cod. Dresd. Db 86. Sie beginnt fol. 200 mit den Definitionen, deren ich 14 zähle, und scheint anfangs dem griechischen Texte ziemlich genau zu folgen, später aber weicht sie mehr und mehr ab. Eine Veröffentlichung dieses Traktates ist dringend zu wünschen."

32) Axel Björnbo, "Die mittelalterlichen lateinischen Übersetzungen aus dem Griechischen auf dem Gebiete der mathematischen Wissenschaften", *Archiv für die Geschichte der Naturwissenschaft und der Technik*, Vol. I (1909), p. 390.

George Valla and Bartholomew Zamberti. Valla gave a partial translation of the *Data* in his voluminous *De expetendis et fugiendis rebus opus* (1501), an encyclopedic collection of Latin translations of Greek learning, science, and art. His translation of the *Data,* which is entitled *De datis* in Chapter 20 of Book XI of the above work, is a part of the translation, which Heiberg observed to be "unbelievably bad" and often "simple replacement of words without taking the meaning into consideration."[33] Zamberti translated the whole of the *Data,* which is included in his Latin version of the *Elements* (1505) as *Euclidis Megarensis philosophi Platonici mathematicique prestantissimi liber datorum.* Menge remarked that this was the first whole Latin translation of the *Data,*[34] but that is obviously not true. Before his translation there existed our medieval Latin version which gave the whole translation of the Greek *Data.* Although Zamberti seems to have consulted Valla's translation published four years before, his rendering is not unfrequently different from Valla's. As a whole, Zamberti's translation is rather better than Valla's. Both these translations seem to be based on Theonine manuscripts.

Following is a parallel presentation of the part on definitions and proposition I from the two translations:

Valla's version	Zamberti's version
Diffinitiones	Diffinitiones
1. Datae magnitudine dicuntur areolae, et lineae, et anguli, quibus aequales possumus exhibere.	1. Data magnitudine dicuntur areae, lineae, et anguli quibus aequalia possumus exhibere.
2. Ratio dari dicitur, cui possumus eandem exhibere.	2. Ratio dari dicitur cui eandem possumus exhibere.
3. Rectilineae figurae specie dari dicuntur, quarum anguli dati sunt ad unum omnes, et rationes laterum ad se invicem datae.	3. Rectilineae figurae specie dari dicuntur, quarum anguli dati sunt ad unum et laterum rationes adinvicem date.

33) Cf. J. L. Heiberg, "Philologische Studien zu griechischen Mathematikern III. Die Handschriften Georg Vallas von griechischen Mathematikern", *Jahrbücher für classische Philologie*, Suppl. Vol. XII (1881), p. 378. "Die griechischen Vorlagen sind sehr flüchtig und ganz willkürlich excerpirt und in wörtlicher, unglaublich schlechter Uebersetzung wiedergegeben, . . . Aber eben diese Schlechtheit der Uebersetzungen, die sich an die Worte halten, ohne viel nach dem Sinn zu fragen, ermöglicht es, die griechischen Vorlagen zu reconstruiren, etc."

34) Cf. Menge's prolegomena, *Euclidis opera omnia*, Vol. VI, p. LVI, "Totum Datorum librum primus Latine Bartholomaeus Zambertus Uenetiis 1505."

4. Positione dari dicuntur puncta et lineae, et anguli, quae eundem semper obtinent locum.	4. Positione dari dicuntur signa, lineae, et anguli, quae eundem semper locum obtinent.
5. Circulus magnitudine dari dicuntur cuius datur quae ex centro magnitudini.	5. Circulus magnitudine dari dicuntur, cuius quae ex centro magnitudine datur.
6. Positione autem et magnitudine circulus dari dicitur cuius datur quidem centrum positione, at quae ex centro magnitudine.	6. Positione magnitudineque circulus dari dicitur, cuius centrum positione datur, ea queque ex centro magnitudine.
7. Sectiones circuli magnitudine dari dicuntur, in quibus anguli dati sunt et bases sectionum magnitudine	7. Segmenta circuli magnitudine dari dicuntur, in quibus anguli dati sunt, et segmentorum bases magnitudine.
8. Positione autem et magnitudine sectiones dari dicuntur, in quibus anguli dati sunt magnitudine, et bases sectionum positione, et magnitudine.	8. Positione et magnitudine segmenta dari dicuntur, in quibus et anguli dati sunt magnitudine, et bases segmentorum positione et magnitudine.
9. Magnitudo magnitudine dato maior est, cum ablato dato, reliquum eidem fuerit aequale.	9. Magnitudo magnitudine dato maior est, quando sublato dato, reliquum eidem aequum fuerit.
10. Magnitudo magnitudine dato minor est cum addito dato totum eidem aequale fuerit.	10. Magnitudo magnitudine dato minor est, quando adiecto dato, totum eidem aequum fuerit.
11. Magnitudo magnitudine dato major est quam sit ratione cum ablato dato reliquum ad idem rationem habuerit datam.	11. Magnitudo magnitudine dato maior est quam in ratione, quando ablato dato, reliquum ad idem rationem datam habuerit.
12. Magnitudo magnitudine dato minor est quam in ratione cum adiecto dato totum ad idem rationem habuerit datam.	12. Magnitudo magnitudine dato minor est quam in ratione, quando apposito dato reliquum ad idem rationem habuerit datam.
13. Producta est quae a dato puncto in positione rectam lineam. (?)	13. Producta est quae a dato signo in positione rectam lineam, acta recta linea in datum angulum, vel datum signum.
14. Producta recta linea in datum punctum acta est, quae dato punto additione reducta recta linea in datum angulum. (?)	14. Reducta quae a dato signo ad positione rectam lineam recta linea in angulo dato acta est.
15. Appositione est quae per da-	16 (15). Appositione est quae per

tum punctum positione recta linea parallelus acta est.

datum signum positione recte linee parallelus acta est.

Propositio I

Magnitudinum datarum ratio ad se invicem data est.

Sint datae magnitudines *a b*. Aio ipsius *a* ad *b* rationem esse datam. Nam quoniam datum est *a*, possibile est ipsi aequale exhiberi. Exhibeatur utique sitque *c*. Rursus quoniam datum est *b*, possibile ipsi aequale exhiberi. Exhibeatur sane sitque *d*. Igitur aequale est *a* quidem ipsi *c* et *b* ipsi *d*. Est igitur ut *a* ad *c* ita *b* ad *d*, et contra, ut a ad *b* ita *c* ad *d*. Ipsius ergo *a* ad *b* ratio est data. Nam idem sibi exhibitum est quae ipsius *c* ad *d*.

Propositio I

Datarum magnitidinum ratio adinvicem datur.

Sint datae magnitudines *a*, *b*. Dico quod ipsius *a* ad *b* ratio data est. Quoniam enim datur *a*, possible est per primam diffinitionem ei aequam exhibere. Exhibeatur et esto *c*. Rursus quoniam data est *b*, possibile est per eandem eidem aequam exhibere. Exhibeatur et esto *d*. Quoniam igitur *a* est equalis ipsi *c* et *b* ipsi *d*, est igitur sicut *a* ad *c* sic est *b* ad *d*. Vicissim per XVI, V elementorum sicut *a* ad *b* sic *c* ad *d*. Ipsius igitur *a* ad *b* ratio data est. Eadem namque eidem exhibetur ipius *c* ad ipsam *d*.

Finally I should say something about the modern translations of the *Data*.

I have already mentioned that Peyrard translated the *Data* into French at the beginning of last century. It seems that this is still the best translation of the *Data* into a modern language. Before that, as Peyrard tells us,[35] Henrion also translated it into French and published it in 1615, although his translation was very bad.

It is often said that Robert Simson made an English translation of this work.[36] His English version of the *Data* in 1756,[37] however, is not really a translation but a free paraphrase. He not only omitted many definitions and propositions which seemed to him inadequate, but also radically changed the order of propositions. He also did not hesitate to insert new propositions and to combine propositions. Consequently, in spite of his intention to restore Euclid's text by correcting Theon's vitiation, his

35) Peyrard, *Op. cit.*, Vol. I, p. xxvii.

36) Heath, *Op. cit.*, Vol. I, p. 8, note 2 etc.

37) *The Elements of Euclid, viz. the first six Books together with the eleventh and twelfth. In this Edition the Errors, by which Theon or others have long ago vitiated these Books, are corrected, and some of Euclid's Demonstrations are restored. Also The Book of Euclid's Data in same manner corrected by Robert Simson.* Glasgow (1756)

version of the *Data* deviated even farther from the original text of Euclid.

While there is repeated reference to Simson's English version of the *Data*, one seldom finds mention of the fact that the English version of Euclid by Isaac Barrow also contains a translation of the *Data* revised by Thomas Haselden.[38] Although this translation was done in the somewhat archaic English, it is more faithful to Euclid's text than Simson's translation.

My English translation, since its Latin original agrees very well with Menge's Greek text, is also fairly faithful to the Greek text. Therefore it will be of use as the only modern English version of the *Data* which is now available.

As far as German versions of the *Data* are concerned, they were made by Schwab and Wurm. Schwab rendered Simson's translation into German in 1780, while Wurm made a direct German translation of the *Data* in 1825. Thaer published a German version of the Data translated from Menge's Greek text in 1962.[39] It should also to be noted that Buchbinder made in 1866 a partial edition of the *Data* based on Peyrard's text with annotations to variants in a Munich manuscripts.[40]

The following diagram will help to grasp at a glance the history of the *Data* which I have traced in this chapter.

38) *Euclid's Elements. The whole Fifteen Books compendiously demonstrated with Archimedes' Theorem of the Sphere and Cylinder investigated by the Method of Indivisibles by Isaac Barrow D. D. To which is annexed Euclid's Data and a brief Treatise of Regular Solids. The whole revised with great care and some hundreds of errors of the former impression corrected by Thomas Haselden.* London (1732)

39) Clemens Thaer, *Die Data von Euklid*, Berlin-Göttingen-Heidelberg, 1962.

40) Cf. *Euclidis Opera Omnia*, Vol. VI, Prolegomena (by Menge) pp. lxi-lxii and also M. Cantor, *Vorlesungen über die Geschichte der Mathematik*, Vol. I, p. 282, note 1.

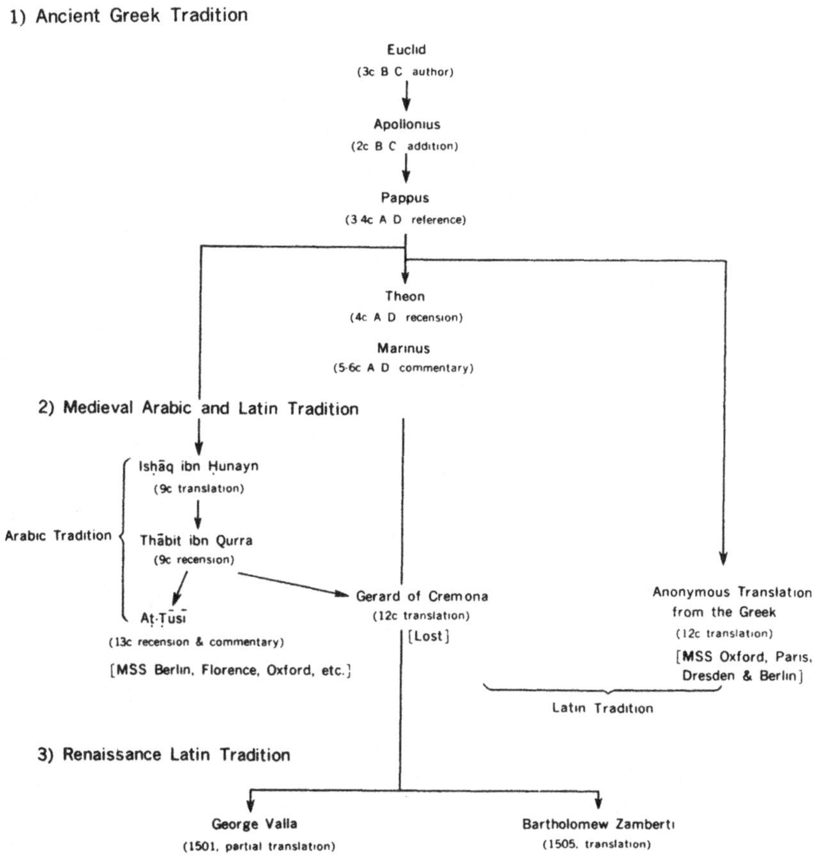

Modern Edition

Greek Text:
- Hardy (1625)
- Barrow (1659)
- Gregory (1703)
- Peyrard (1818)
- Buchbinder (1866)
- Menge (1896)

Latin Version:
- Valla (1501)
- Zamberti (1505)
- Menge (1896)
- Ito (1980)

Modern Translation

- Henrion (F. 1615)
- Haselden (E. 1732)
- Simson (E. 1756)
- Schwab (G. 1780)
- Peyrard (F. 1818)
- Wurm (G. 1825)
- Thaer (G. 1962)
- Ito (E. 1980)

CHAPTER 3

AN ANALYSIS OF THE MEDIEVAL LATIN TRANSLATION FROM GREEK OF THE *DATA*

The four medieval manuscripts employed for my edition and translation of the *Data* contain substantially the same text of this Latin translation from the Greek. That this Latin translation of the *Data* was made from the Greek is obvious from the following:

1. the existence of the direct transliteration of Greek words, like *catigmeni, anigmeni, parathesi, cathetus, orthogonius* etc.

2. the existence of an exact correspondence in particles and conjunctions used in this Latin translation and the Greek text. The following main correspondences should be noted:

ἀλλά	= sed, at	ἄρα, οὖν	= ergo, igitur
γάρ	= enim	δέ	= vero, -que, atqui at, et, autem
δή	= autem, et, vero	εἰ, ἐάν	= si
ἐπεί	= quoniam, quia	ἤ . . . ἤ	= vel . . . vel
καί	= et, etiam, atqui -que, quoque	μέν . . . δέ	= quidem . . . vero, autem
μή, οὐ	= non, minime	ὅτι	= quod, quoniam
τέ . . . τέ	= et . . . et	πάλιν	= rursum
ὡς . . . οὕτως	= sicut . . . ita, quemadmodum . . . ita	ὥστε	= quare

3. the use of the order of Greek letters in the diagrams, like abgdez (αβγδεζ).

4. the same order of words which is often found between the Greek original and the Latin version, like ἅ τὸν αὐτὸν ἀεὶ τόπον ἐπέχει = que eundem semper locum optinent.

In this chapter, I shall investigate two aspects of this translation from Greek of the *Data*: first, its date and place of translation; second, its possible translator. Then I shall make a comparative examination of the four manuscripts.

(i) Date and Place of the Translation

Since our manuscripts give no direct information about the date and

place of this Latin translation, we must first seek out likely possibilities from the general circumstances of the cultural history of medieval Europe.

As is well known, the positive assimilation and recovery by the Latin West of Greek science and civilization first took place in the 12th century, primarily through translations made from Arabic and Greek.

The rediscovery of Greek learning through Arabic intermediaries came about mainly in Spain and particularly but not exclusively in the city of Toledo, where Archbishop Raymond urged Arabic study by establishing a school for Latin translation of Arabic versions and putting it in the charge of his archdeacon Domingo Gundisalvo. Such noted names as Adelard of Bath, Plato of Tivoli, Robert of Chester, Hermann of Carinthia, Gerard of Cremona, and Michael Scot participated in this cultural activity of rendering Latin versions of Greek scientific works from Arabic.

But, since it is manifest that our Latin translation came not from Arabic but from Greek, we need not consider this Arabic element of the "12th-century Renaissance." For our translation we must consider rather the other element, namely, the efforts to bring Greek works into Latin immediately from Greek.

Translation from Greek itself took place in the 12th century in two areas—in northern Italy and in the Norman kingdom of southern Italy, and especially in Sicily.

Since some cities in northern Italy, like Venice and Pisa, kept close commercial ties with Constantinople and since Sicily was long a part of the Byzantine Empire, these regions still retained Greek influence and had not lost contact with the East. In Sicily, in particular, Greek, Latin, and Arabic people lived side by side in peace and tolerance, and the languages of all three groups were actively in use there. Such circumstances undoubtedly favored Italy as a place of cultural transmission.

Henricus Aristippus and Eugene of Palermo emerged as the principal translators in Sicily, while James of Venice, Burgundio of Pisa, and Moses of Bergamo were the most active translators in northern Italy.

When we compare the Latin translations from Greek made in northern Italy with those made in Sicily, we note a remarkable difference in character. The translators in northern Italy were mainly concerned with translation of 1) the works of Aristotle such as the *Topics*, the *Prior* and *Posterior Analytics*, the *Elenchi* and *Physics*, etc. and 2) "theological" works like those of Basil, Chrysostom, and John of Damascus.[41)]

41) Charles H. Haskins, *Studies in the History of Medieval Science*. Cambridge, Mass., 2nd ed. (1927), p. 144 seq. & Chap. X.

In contrast with this "philosophical-theological" tendency of the translators in northern Italy, the "mathematical-scientific" bent of the Sicilian school is quite conspicuous. We know that in Sicily Latin translations were produced of such scientific works as Ptolemy's *Almagest* and his *Optics*, Euclid's *Optics* and *Catoptrics*, Proclus' *On Motion* and Hero's *Pneumatics*.[42] These works were all translated from Greek with the single exception of Ptolemy's *Optics*, which was translated from Arabic.

In view of the dominant mathematical-scientific character of the Sicilian translation, it is quite probable that our Latin translation of Euclid's *Data* was also made in Sicily. This possibility is strengthened by the fact that Euclid's other minor works, like the *Optics* and *Catoptrics*, were translated there.

Thus, among the three major areas of the "medieval renaissance," Spain, northern Italy, and Sicily, the first was the center of Latin translation from "Arabic" rather than Greek, while in northern Italy there is no evidence that any "mathematical" work was translated from Greek into Latin. Thus we are left with the island of Sicily, where many Greek mathematical scientific works, including Euclid's, were translated into Latin from Greek.

The conjecture that our translation of the *Data* was made in Sicily will also be strengthened by our discussion of the translator in the next section.

As regards the date of this Latin translation of the *Data*, we might suppose that the translation came into existence in the 12th century, because the other above-mentioned translations of Euclidian minor works in Sicily were produced in that century. It does not seem likely that the translation of the *Data* is an exception.

One might also argue that since the translation is very faithful, even painfully literal, it expresses a characteristic feature of 12th-century translation, although the occasional persistence of literal translations into the 13th century makes such an argument doubtful. Still, one of the general features of the 12th-century translations is that they are excessively slavish in translation.[43] And indeed the Latin translation of the *Data* is so literal that almost every Greek word is represented by a Latin equivalent, even including the particles peculiar to Greek like μέν . . . δέ, and the order of words often conforms to the Greek rather than the Latin mode. This

42) *Ibid.*, p. 143 & Chap. IX.

43) Literal translations on occasion come not so much from the linguistic ignorance of the 12th-century translators as from their belief that the text they were rendering was an authoritative one whose words should not be trifled with. Cf. Burgundio's preface to Chrysostom's work in Ch. H. Haskins, *Op. cit.*, pp. 151–2.

method of translation *de verbo ad verbum* predisposes us to accept a 12th-century origin of the *Data*, if other evidence seems to point to that period.

Thus, following our argument to this point, we can speculate that this translation was made in the 12th century. And if my conjecture is valid for identifying the translator, it must have been completed *not later* than 1160.

(ii) Identification of the Translator

Generally speaking, it is a difficult task to identify the translators of the 12th century, since so many manuscripts of this century contain no reference to the translator.

Since this is true also of the manuscripts of the Latin translation of the *Data*, we must attempt to identify him by comparing our Latin translation with other available Latin translations from Greek of the same period. This comparison is best made by seeking out distinctive peculiarities of translation, *e.g.*, by paying special attention to some crucial particles.

On this point, Minio-Paluello's many studies on early medieval Latin translations from Greek can serve as the point of departure for our investigation, for he has made detailed comparisons of the Greek texts of a number of works (primarily of Aristotle) with their early medieval Latin vessions.[44)]

Using his studies, I checked the translating techniques of such principal translators of the 12th and 13th century as Burgundio of Pisa, Bartholomew of Messina, Henricus Aristippus, Robert Grosseteste, William of

44) Minio-Paluello's articles I could use are the following:

1. Lorenzo Minio-Paluello, "Henri Aristippe, Guillaume de Moerbeke et les traductions latines médiévales des «Météorologique» et du «De generatione et corruptione» d'Aristote", *Revue philosophique de Louvain*, Vol. XLV (1947), pp. 206–235.

2. ———, "Les 'trois rédactions' de la traduction médiévale gréco-latine du «De generatione et corruptione» d'Aristote", *Ibid.*, Vol. XLVIII (1950), pp. 247–259.

3. ———, "Guglielmo di Moerbeke, Traduttore della Poetica di Aristotele", *Rivista di Filosofia neo-scolastica*, Anno XXXIX—Fasc. I (1947), pp. 3–19.

4. ———, "Note sull'Aristotele latino medievale, I-III", *Ibid.*, Anno XLII—Fasc. III (1950), pp. 1–16.

5. ———, "Note sull'Aristotele latino medievale, V-VII", *Ibid.*, Anno XLIV (1952), pp. 389–495.

6. ———, "Note sull'Aristotele latino medievale, VIII-IX", *Ibid.*, Anno XLVI—Fasc. III (1954), pp. 211–231.

7. ———, "Le texte du «De anima» d'Aristote, La traduction latine avant 1500", *Autour d'Aristote* offert à Monseigneur A. Mansion. Louvain (1955), pp. 217–243.

Moerbeke, and two other anonymous translators of Aristotle's works. I could not find, however, any convincing agreement in translating techniques between our Latin translator of the *Data* and those analyzed by Minio-Paluello. A crucial point which distinguishes our translator from them is that he never used the Latin particles *utique* or *itaque* for the Greek δή but employed *autem, et, vero* for it.[45] On this point, then, his translation is crucially different from those of Burgundio, Bartholomew, Aristippus, Grosseteste, Moerbeke, and the other two anonymous translators,[46] although it agrees with each of them in the rendering of some other particles. To be sure, the work being translated is mathematical rather than philosophical, but presumably the manner of translating simple particles ought to carry over from one type of work to another.

After I failed to find our translator among those whom Minio-Paluello examined, I searched for him anew on the basis of the conclusion suggested in the preceding section, that the Latin translation of the *Data* was made in Sicily in the 12th century. In this regard, Charles Homer Haskins' study of Latin translation from the Greek in Sicily was helpful.[47] Hence I have depended on his study in my examination of the major Sicilian translators.

It is now well known that in 12th-century Norman Sicily there were two principal translators who were also members of the royal administration under William I. The first was Henricus Aristippus, to whom is attributed the lost Latin versions of Gregory Nazianzen and Diogenes Laertius. The second was Eugene of Palermo, known as the transmitter to the West of two pieces of Oriental literature, the prophecy of the Erythraean Sibyl and the Sanscrit fable of Kalila and Dimna.

Aristippus is known to have been the first Latin translator of Plato's *Meno* and *Phaedo* as well as of Aristotle's *On Generation and Corruption* and *Meteorology* (Book IV). But it is clear from my examination of his

45) The exact times of use of Latin words for δή in our translation of the *Data* are as follows: autem (14), et (10), vero (7), at (1), enim (1) and *om.* (1).

46) Their specific translations of the Greek particle δή are as follows:

Burgundio, *utique*; Bartholomew, *utique*; Aristippus, *utique*; Grosseteste, *utique*; Moerbeke, *itaque* (very often) or *utique*; anonymous translator of the *Posterior Analytics, om.* or *igitur*; anonymous translator of the *De somnio et vigilia, quidem, quare* or *itaque*.

Cf. the above-cited papers of Minio-Paluello, especially, articles 1 & 4.

47) Charles H. Haskins, *Studies in the History of Medieval Science*, Cambridge, Mass., 2nd ed. (1927), Chap. VIII & IX.

Also cf. his other work, *The Renaissance of the 12th Century*, Cleveland—New York: Meridian Books (1957), Chap. IX.

particular translating manner found in the articles of Minio-Paluello that he is not the translator of the *Data*.[48] While the second translator—Eugene—is indeed important in the history of science, because he gave a Latin version from Arabic of Ptolemy's *Optics*, which otherwise would have been lost, there is also no evidence that he translated the *Data* of Euclid, though Haskins tells us "it is not surprising to learn that he had at hand the Greek text of Euclid's *Data, Optica* and *Catoptrica*, as well as the treatise of Proclus on mechanics."[49]

In addition to Aristippus and Eugene, however, there exists a third figure whose activity has already been described by Heiberg and Haskins.[50] He is the anonymous translator of Ptolemy's *Almagest*, the manuscripts of which were discovered at the beginning of this century by Lockwood, Björnbo, and Haskins.[51] Some of these manuscripts contain a fairly long preface by the translator, which is very useful. Haskins quoted it in his work on medieval science.[52]

In this preface, dedicated to his teacher of mathematics, the translator says that "as he was laboring over the study of medicine at Salerno, he learned that a copy of Ptolemy's great treatise had been brought from Constantinople to Palermo, as a present from the Greek emperor, by an ambassador of the Sicilian king. This emissary, by name Aristippus

48) Besides his use of *utique* for δή, another point which distinguishes Aristippus from our translator is the former's use of *quapropter* for ὥστε, which is never found in our translation of the *Data*.

49) Haskins, *Studies in the History of Medieval Science*, p. 171.

50) Cf. the following articles:

J. L. Heiberg, "Eine mittelalterliche Übersetzung der Syntaxis des Ptolemaios", *Hermes*, Vol. XLV (1910).

Heiberg, "Noch einmal die mittelalterliche Ptolemaios-Übersetzung", *Ibid.*, Vol. XLVI (1911).

Ch. H. Haskins & P. Lockwood, "The Sicilian Translators of the Twelfth Century and the First Version of Ptolemy's *Almagest*", *Harvard Studies in Classical Philology*, Vol. XXI (1910).

Haskins, "Further Notes on Sicilian Translators of the Twelfth Century", *Ibid*, Vol. XXIII (1912).

Also see A. Pelzer's paper in *Archivum Franciscanum Historicum*, Vol. XII (1919).

51) The MSS of this Latin translation of Ptolemy's *Almagest* which have been discovered are the following:

1. MS Vat. lat. 2056; 2. MS Flor. Bibliotheca Nazionale, Conventi Soppressi, A. 5.2654; 3. MS Vat. Pal. lat. 1371; 4. MS Wolfenbüttel, Gud. lat. 147, f. 2. (preface only).

Cf. Haskins's papers in *Harvard Studies in Classical Philology*, Vol. XXI (1910), pp. 75–102 & Vol. XXIII (1912) pp. 155–166.

52) Haskins, *Studies in the History of Medieval Science*, pp. 191–3.

[none other than Henricus Aristippus], he set out to seek, and braving the terrors of Scylla and Charybdis and the fiery streams of Etna, he found him at Pergusa, near the fount, engaged, not without danger, in investigating the marvels of Etna." But the mathematical-astronomical knowlledge of this Salernitan scholar "was not sufficient to permit his attempting at once the translation of the book which he had sought, even if there had been no other obstacles in the way, and already familiar with Greek, he applied himself diligently to the preliminary study of Euclid's *Data, Optics* and *Catoptrics* and *De motu* of Proclus. When ready to attack the *Almagest*, he had the good fortune to find a friendly expositor in Eugene, a man most skilled in Greek and Arabic and not unfamiliar with Latin, and succeeded, contrary to the desire of an ill-tempered man, in turning the works into Latin."[53]

In this preface, we should turn our attention to the statement "already very diligently instructed in the Greek language, first of all, I made a preliminary study of Euclid's *Data*, *Optics* and *Catoptrics* and also Proclus' *Elementatio physica*." (grecis ego litteris diligentissime preinstructus, primo quidem in Euclidis Dedomenis, Opticis, et Catoptricis, Physicaque Procli Elementatione prelusi.)[54]

What the verb *preludo* means in the above sentence is not clear. But it seems to be possible to interpret it as implying that the preliminary understanding was not just one of simply studying the works but also of rendering them into Latin.[55]

This conjecture is supported by the fact that there exist manuscripts of a Latin translation of Proclus' *De motu* or *Elementatio physica*, whose translator Haskins identified with the translator of the *Almagest* mentioned above.[56] Haskins made this identification on the basis of close resemblances in translating technique.

If this anonymous Salernitan scholar versed in mathematics did translate from Greek into Latin not only the *Almagest* but also Proclus' *Ele-*

53) Haskins, *Op. cit.*, pp. 159–160.

54) Haskins, *Op. cit.*, p. 191.

55) Actually, Heiberg interpreted this word *prelusi* to imply "to have translated". See J. L. Heiberg, "Noch einmal die mittelalterliche Ptolemaios-Übersetzungen", *Hermes*, Vol. XLVI (1911), p. 209.

"Da Aristippus wegen mangelnde Kenntnisse der Astronomie nicht selbst die Syntaxis übersetzen konnten, überliess er diese Arbeit unserem Mann, *der schon die griechische Sprache beherrschte und als Vorübung Euklids Data, Optik, Katoptrik nebst der Στοιχείωσις φυσική ἢ περὶ κινήσεως des Proklos übersetzte.*"

But Heiberg did not give any justification for this interpretation.

56) Haskins, *Op. cit.*, pp. 180–181.

mentatio mentioned in the preface to his translation of the *Almagest*, it is reasonable to assume that he also rendered into Latin the other works which he listed along with the *Elementatio* in the preface—namely, Euclid's *Data, Optics*, and *Catoptrics.*

Consequently, I compared, first of all, the translating technique used in the *Data* with that of Proclus' *Elementatio*, the full Latin text of which has been published together with the Greek original by H. Boese.[57] As the result of this comparison, I found that there was a remarkable agreement between them. They accord very well in translations of almost all connective terms. The correspondence of conjunctives in the *Elementatio* is as follows:

ἀλλά =	sed, verum	*ἄρα, οὖν* =	ergo, igitur
γάρ =	enim	*δέ* =	autem, vero, et, -que
δή =	autem, vero	*εἰ, ἐάν* =	si
ἐπεί =	quoniam	*ἤ . . . ἤ* =	vel . . . vel
καί =	et, etiam,- que› vero	*μή* =	non
μέν . . . δέ =	quidem . . . vero or autem	*ὅτι* =	quod, quoniam
οὐ =	minime, non	*πάλιν* =	rursum, rursus
οὕτως =	ita	*ὥστε* =	quare

If one compares these correspondences with those in the *Data* as given at the beginning of this chapter, the agreement between them is remarkable. They agree not only in the correspondence of almost all connective terms but in the usage of other words like *ἄμφω* = ambo, *ἀλλήλων* = ad invicem, *ἄλλως* = aliter, *ἐφεξῆς* = deinceps, *ἐπ' ευθεῖας* = in directo, *μέγεθος* = magnitudo, quantitas etc.

In the following I shall give from Boese's text a few passages of Proclus' *Elementatio physica*, its Greek original and Latin translation, as examples for comparison. The underlined correspondences of particular terms between Greek and Latin are found exactly in the same way in the translation of the *Data*. Especially note the correspondence *δή* = *autem, vero,* which is peculiar to the translator of the *Data* and distinguishes his work from that of the other translators in the same period, as I mentioned.

ELEMENTATIO PHYSICA

Δύο ἀμερῆ συνεχὲς οὐδὲν ποιήσει.	2. Duo individua continuum nichil faciunt.

57) Hermut Boese, *Die mittelalterliche Übersetzung der ΣΤΟΙΧΕΙΩΣΙΣ ΦΥΣΙΚΗ des Proclus*, Berlin, 1958.

Εἰ γὰρ δυνατόν, ἔστω δύο ἀμερῆ τὰ ΑΒ καὶ ποιείτω συνεχὲς τὸ ἐξ ἀμφοῖν. ἀλλὰ πάντα τὰ συνεχῆ ἅπτεται πρότερον· τὰ ἄρα ΑΒ ἅπτεται ἀλλήλων ἀμερῆ ὄντα, ὅπερ ἀδύνατον.	Si enim possibile, sint duo individua A et B et faciant continuum quod ex ambobus. Sed omnia continua contingunt se prius; ergo A B contingunt se ad invicem, individua existentia, quod est impossibile.
῎Αλλως. Εἰ ἔστι συνεχὲς ἐκ ΑΒ ἀμερῶν, ἢ ὅλον ὅλου ἅπτεται τὸ Α τοῦ Β ἢ ὅλον μέρους ἢ μέρει μέρους. ἀλλ' εἰ μὲν ὅλον μέρους ἢ μέρει μέρους, οὐκ ἔσται ἀμερῆ τὰ ΑΒ, εἰ δὲ ὅλον ὅλου ἅπτοιτο, οὐκ ἔσται συνεχές, ἀλλ' ἐφαρμόσει μόνον· εἰ οὖν οὐκ ἦν τὸ Α συνεχὲς, οὐδὲ τὸ Β μετὰ τοῦ Α ἔσται συνσχὲς ὅλον ὅλου ἁπτόμενον.	Aliter: Si est continuum ex A B individuis, vel totum totum contingit A B vel totum partem vel partes partem. Sed si totum partem vel partes partem, non erunt individua A B. Si vero totum totum contingat, non erunt individua, sed supponetur tantum. Si ergo non erat A continuum, nec vero B cum A erunt continuum, totum totum contingens.
Δύο ἀμερῆ ἐφεξῆς ἀλλήλοις οὐκ ἔστιν.	4. Duo individua deinceps ad invicem non sunt.
῎Εστω γὰρ δύο ἀμερῆ τὰ ΑΒ· λέγω ὅτι οὐκ ἔσται ἐφεξῆς τὸ Α τῷ Β. ἐπεὶ γὰρ δέδεικται, ὅτι δύο ἀμερῶν τὸ μεταξὺ συνεχές ἐστιν, ἔστω δὴ τὸ μεταξὺ αὐτῶν τὸ ΓΔ καὶ διῃρήσθω κατὰ τὸ Ε· τὸ Ε ἄρα ἀμερές ἐστι μεταξὺ ὂν τῶν ΑΒ· ἐφεξῆς δὲ ἦν, ὧν μηδὲν μεταξὺ ὁμογενές· οὐκ ἄρα τὸ Α καὶ τὸ Β ἐφεξῆς ἐστιν.	Sint enim duo individua A B; dico quod non erit deinceps A ei quod est B. Quoniam enim demonstratum est, quod duorum individuorum intermedium continuum est, sit autem intermedium eorum G D et dividatur ad punctum E. Erit ergo E individuum intermedium existens eorum que sunt AB. Deinceps autem erant, quorum nichil intermedium omogenes. Non ergo A et B deinceps sunt.

Since we have seen that the Latin translation of the *Data* accords remarkably with the Latin translation of Proclus' *Elementatio* in almost all characteristic ways of turning words, I suspect that the medieval Latin translation of the *Data* was made by the same person as the translator of the *Elementatio* and thus of the *Almagest* as well, for Haskins and Boese have already identified the translator of Proclus' *Elementatio* with that of the *Almagest*.[58)]

On the same reasoning, I thought it was quite probable that this translator made the Latin versions of the *Optics* and *Catoptrics* of Euclid, which also were listed in the preface of the translation of the *Almagest*. Hence

58) Cf. Haskins, *Op. cit.*, p. 181 and H. Boese, *Op. cit.*, p. 19.

I checked a manuscript of the Latin translations of these two minor works attributed to Euclid, which is MS Oxford, Bodl., Auct. F.5.28, 13c. On folios 57r-64r (cat. pag. 17r-24r) it contains Euclid's *Optics* under the title *De visu liber*[59]; and on folios 64r-69r (cat. pag. 24r-29r) it contains Euclid's *Catoptrics*, entitled *De speculis liber*.

By comparative examination, I found close agreement in translating techniques between the Latin translations of the *Data* and the *Elementatio* and those of the *Optics* and *Catoptrics*. Actually, our Latin translation of the *Data* accords quite well with those of the *Optics* and *Catoptrics* not only in translation of almost all connective terms but also in the usage of many common fundamental words. Thus we can say with certainty that our translator of the *Data* also made the Latin version of the *Optics* and *Catoptrics* of Euclid.

For the reader's reference, I shall present here sample passages of these Latin translations of the *Optics* and *Catoptrics* with their Greek original given in Heiberg's text. The identity or similarity of translation of the *Data* with them is very palpable. The underlined correspondences are found also in the translation of the *Data*. Especially note here again the correspondence δή = *autem, vero*, which is peculiar to the translator of the *Data*.

OPTICS

Τὰ ὀρθογώνια μεγέθη ἐξ ἀποστήματος ὁρώμενα περιφερῆ φαίνεται.	IX. Rectangule magnitudines e distantia vise periferie apparent.
Ἔστω γὰρ ὀρθογώνιον τὸ ΒΓ ἑστὼς μετέωρον ἐξ ἀποστήματος ὁρώμενον. οὐκοῦν, ἐπεὶ ἕκαστον τῶν ὁρωμένων ἔχει τι μῆκος ἀποστήματος, οὗ γενόμενον οὐκέτι ὁρᾶται, ἡ μὲν Γ ἄρα γωνία οὐχ ὁρᾶται, τὰ δὲ Δ, Ζ σημεῖα μόνον φαίνεται. ὁμοίως καὶ ἐφ' ἑκάστης τῶν λοιπῶν γωνιῶν τοῦτο συμβήσεται ὥστε ὅλον περιφερὲς φανήσεται.	Esto enim rectangulum BG existens elevatum e distantia visum. Igitur quoniam unumquodque visorum habet longitudinem distantie, qua facta non iam videtur, angulus G quidem non videtur, puncta vero D, Z tantum apparent. Similiter et in unoquoque reliquorum angulorum hoc continget. Quare totum periferium apparebit.
Τῶν εἰς τοὔμπροσθεν μῆκος ἐχόντων τὰ μὲν ἐν τοῖς δεξιοῖς εἰς τὰ ἀριστερὰ δοκεῖ παρῆχθαι, τὰ δὲ ἐν τοῖς ἀριστεροῖς εἰς τὰ δεξιά.	XII. In ante habentium longitudinem que quidem in dextris, in sinistra, que vero in sinistris in dextra educi videntur.

59) Heiberg published an edition of this medieval Latin version of the *Optics* based on one manuscripts in *Euclidis Opera Omnia*, Vol. VII. Also consult the more critical edition in Theisen's thesis (see the bibliography).

Ἔστω δύο ὁρώμενα μεγέθη, τὰ ΑΒ, ΓΔ, ὄμμα δὲ ἔστω τὸ Ε, ἀφ' οὗ προσπιπτέτωσαν ἀκτῖνες αἱ ΕΘ, ΕΚ, ΕΑ, ΕΖ, ΕΗ, ΕΓ. λέγω, ὅτι αἱ μὲν ΕΖ, ΕΗ, ΕΓ δοκοῦσιν εἰς τὰ ἀριστερὰ μετῆχθαι, αἱ δὲ ΕΘ, ΕΚ, ΕΑ εἰς τὰ δεξιά. ἐπεὶ γὰρ ἡ ΕΖ τῆς ΕΗ ἐστι δεξιωτέρα, ἡ δὲ ΕΗ τῆς ΕΓ, ἐντεῦθεν ἄρα ἡ ΕΓ τῆς ΕΗ δοκεῖ εἰς τὰ ἀριστερὰ μετῆχθαι, ἡ δὲ ΗΕ τῆς ΕΖ. ὁμοίως καὶ αἱ ΕΚ, ΕΑ, ΕΘ δοκοῦσιν εἰς τὰ δεξιά μετῆχθαι.

Sint due magnitudines conspecte AB et DG, sitque oculis E, a quo accidant radii ET et EK et EA, item EZ, EI et EG. Dico quod EZ et EI et EG videntur in sinistra protracte, ET vero et EK et EA in dextra. Quoniam enim EZ dexterior est quam EI, atque EI quam EG, inde ergo EG ab EI videtur in sinistra tracta, EI *vero* ab EZ. Similiter et EK, EA, ET videtur in dextra tracte.

CATOPTRICS

Δυνατόν ἐστι διὰ πλειόνων ἐνόπτρων ἐπιπέδων ἰδεῖν τὸ αὐτό.

Ἔστω, ὃ δεῖ ὀφθῆναι, τὸ Α, ὄμμα δὲ τὸ Β, ἔνοπτρα δὲ τρία τὰ ΓΔ, ΔΕ, ΕΖ. ἤχθω δὴ κάθετος ἀπὸ τοῦ Β ἐπὶ τὸ ΓΔ ἔνοπτρον ἡ ΒΓ, ἴση δὲ ἡ ΒΓ τῇ ΓΖ, καὶ πάλιν ἀπὸ τοῦ Α ἐπὶ τὸ ΕΖ κάθετος ἡ ΑΖ, καὶ τῇ ΑΖ ἴση ἡ ΖΘ, καὶ ἀπὸ τοῦ Θ ἐπὶ τὸ ΔΕ ἔνοπτρον κάθετος ἤχθω ἡ ΘΚ, καὶ ἔστω τῇ ΘΚ ἴση ἡ ΚΛ, καὶ ἀπὸ τοῦ Λ ἐπὶ τὸ Σ ἐπεζεύχθω ἡ ΛΜΞΣ, ἀπὸ δὲ τοῦ Μ ἐπὶ τὸ Θ ἡ ΜΡΘ, ἐπεζεύχθωσαν δὲ καὶ αἱ ΑΡ, ΒΞ. ἐπεὶ οὖν ἴση ἐστὶν ἡ ΒΓ τῇ ΓΣ, καὶ ὀρθαὶ αἱ πρὸς τῷ Γ γωνίαι, δύο δὴ αἱ ΒΓ, ΓΦ δυσὶ ταῖς ΣΓ, ΓΦ ἴσαι εἰσὶν ἑκατέρα ἑκατέρᾳ, καὶ γωνία ἡ ὑπὸ ΒΓΦ ὀρθὴ οὖσα γωνίᾳ τῇ ὑπὸ ΣΓΦ ὀρθῇ οὔσῃ ἴση ἐστίν, καὶ αἱ λοιπαὶ γωνίαι ταῖς λοιπαῖς γωνίαις ἴσαι ἔσονται, ὑφ' ἃς αἱ ἴσαι πλευραὶ ὑποτείνουσιν, ἡ μὲν πρὸς τῷ Β γωνία τῇ πρὸς τῷ Σ, ἡ δὲ Ξ γωνία τῇ Τ. ἀλλ' ἡ Τ τῇ Ν ἐστιν ἴση· κατὰ κορυφὴν γάρ· ὥστε ἴση ἐστὶ καὶ ἡ Ν γωνία τῇ Ξ. ἡ ἄρα ΒΞ ὄψις ἀνακλασθήσεται ἐπὶ τὸ Μ.

XIII. Possibile est per plura specula plana videri idem.

Esto quod oportet videri A, oculus vero B, specula tria GD et DE et EZ. Trahatur autem cathetus ab B super GD speculum scilicet BG. Equalis vero sit BG ei que est GS, et rursum super EZ cathetus AZ, et que est AZ equalis ZT, et a T super DE speculum trahatur cathetus TK, et esto KT recte equalis KL, et ab L super S trahatur LMNXS, ab M vero super T recta MRT trahatur. Trahantur vero et AR ad punctum contactus MTR et BX ad N. Quoniam ergo equalis est BG recta ei que GS, et recti ad G anguli, due vero BG et GF duabus SG et GF equales sunt utraque utrique, et angulus BGF rectus existens angulo SGF recto existenti equalis est,. et reliqui ergo reliquis angulis equales sunt, quibus equalia latera subtenduntur,ille quidem qui ad B angulus ei qui ad S, et qui ad X ei qui ad C. Sed C ei qui est N equalis est, ad vertices enim sunt. Quare et equalis est N angulus angulo X. Visus ergo BX refringitur super M.

Haskins, who did not consult the Latin manuscripts of the *Data, Optics*, and *Catoptrics*, knew nothing about the relation of the Latin translations of these Euclidian works to those of the *Almagest* and *Elementatio*. Although he supposed that the *Data*, *Optics*, and *Catoptrics* had been translated in Sicily, he was rather negative about the relation between the Latin translations of these works and the *Almagest*, as it is shown when he said "the mention of Euclid's *Data, Optics* and *Catoptrics* helps to connect the Latin translations of these works likewise with the Sicilian school, *if not with the translator of the Almagest himself*"[60] (emphasis mine).

However, since we have attested that the four works mentioned in the preface to the Latin translation of the *Almagest*, namely, the *Data, Optics, Catoptrics*, and Proclus' *Elementatio*, were translated by the same person, we must deduce that the author of these Latin versions also translated the *Almagest*, if we admit the conclusion drawn by Haskins and Boese on the identity of the translator of the *Elementatio* with that of the *Almagest*.

Although not thoroughly, I was able to check the translating technique of connective terms in the Latin translation of the *Almagest* against the list of correspondence of particles and conjunctions given in Boese's work as the result of his examination of the first seven Books of the *Almagest* translation.

The correspondences of main connective terms is as follows:[61]

ἀλλά = sed, verum	γὰρ = enim
δέ = vero, autem, verum	δή = ergo, utique, quidem
ἐπεί = quoniam	ἤ . . . ἤ = vel . . . vel
καί = et, -que, quoque, atque	μέν = quidem
ὅτι = quoniam	οὐ, μή = non, minime
οὖν = igitur	οὕτως = ita, sic
πάλιν = rursus, rursum	ὥστε = quare

First of all, we should notice that the correspondences are quite similar to those in the translation of the *Data* and *Elementatio*. At the same time, however, we should note that there are two major divergences from them.

The one is that the translator of the *Almagest* uses almost exclusively *quoniam*, not *quod*, for the Greek conjunctive ὅτι. But this difference can be explained without much difficulty, when we understand that in the translation of the *Data* the translator would avoid confusion by preferring

60) Haskins, *Op. cit.*, p. 178.

61) H. Boese, *Op. cit.*, pp. 17–18. I omitted the Latin terms which are very rarely used.

quod for ὅτι in order to reserve *quoniam* for ἐπεί in the same proposition. Actually, he preserves his habit of translating ὅτι as *quoniam* in the translation of some propositions of the *Data* (Prop. 64, 66, 67, 91, 92, 93). We can say the same about the translation of Proclus' *Elementatio*.

The other divergence is the translation of δή in the *Almagest* as *ergo, utique*, not as *autem, vero*. This difference is really remarkable. Faced with this substantial divergence, we can follow one of two lines of thought. One is that we suppose that the translator changed his notion of δή in the translation of the *Almagest* after having finished the preliminary translations of the *Data, Optics, Catoptrics*, and *Elementatio*, on the grounds 1) that the preface to the *Almagest* translation identifies the translator of the *Almagest* as that of the other four works and 2) that the translation of the *Almagest* agrees fairly well with those of the other works in other points.[62] Actually, in spite of such divergences Boese has concluded the identity of the translator of the *Almagest* with that of the *Elementatio* on the basis of his examination. If he is right in this conclusion, it must follow that the same scholar made the Latin translation from Greek of at least three works of Euclid, *Elementatio physica* of Proclus and the *Almagest* of Ptolemy, since it is attested by my investigation that the *Elementatio, Data, Optics*, and *Catoptrics* were translated by the same person.

If one feels this way rather expediential, the second approach is to regard the translator of the *Almagest* as a person different from that of the *Data, Optics, Catoptrics*, and Proclus' *Elementatio physica*. This is the way taken by John Murdoch, who has recently discovered two manuscripts of Latin translation of the *Elements* from the Greek——MS Paris Bibl. Nat. Fonds, lat. 7373. 13c, and MS Florence. Bibl. Naz. Cent. Fondo Conventi Soppressi C I 448, 14c——and published an excellent study on it.[63] After minutely checking the translating techniques of the *Elements, Almagest, Data, Optics* and *Catoptrics*, he concluded 1) that the anonymous translator of the *Almagest* and that of the *Elements* are identical and 2) that these two works were not translated by the same person as the *Data, Optics*, and *Catoptrics*. He gives the same divergences as I mentioned above as the most important evidence differentiating the former two works from the latter three, namely, the fact 1) that both consistently

62) See H. Boese, *Op. cit.*, pp. 16–20.
"Es dürfte deshalb kaum einem Zweifel unterliegen, dass die uns erhaltene Übersetzung der Elementatio physica tatsächlich mit derjenigen gleichzusetzen ist, die wir als ein Werk des anonymen Ptolemäusübersetzers kennengelernt haben." (p. 19)

63) J. Murdoch, "Euclides Graeco-Latinus", *Harvard Studies in Classical Philology*, Vol. 71 (1967).

render ὅτι by *quoniam*, while the three minor works choose *quod* as often as *quoniam* and 2) that the former two give *ergo* as translation of δή in almost cases, while the latter three rendered it as *autem, vero*, and *et*. He also adds two salient differences: (a) both the *Elements* and the *Almagest* employ *copulare* for ἐπιζευγνύναι, while the other three works use *coniungere* and other words; b) the two longer works show a marked preference for *scribere* in rendering γράφειν, while the other three use *describere*.[64]

I completed my Wisconsin thesis in 1962, four years before Murdoch's article appeared. Still, now that I have read his paper, it seems to me that his arguments are very cogent. The truth might be that there existed two groups of Latin versions of scientific works prepared by two different persons in 12th-century Sicily. If the translator of the *Almagest* is not the same as that of the *Data* and other works, *preludo* in the preface of the *Almagest* does not mean a translation but simply a preliminary study. In that case, the translator of the *Almagest* probably read the latter works in a Latin translation which had already been made by the other author.

At the same time, however, when I had an occasion to read a copy of the Paris manuscripts of the Latin *Elements*, I could not help being impressed by the remarkable similarity as a whole in style of translation between the *Elements* and the *Data*. Faced with this impression I am tempted to believe that they are the work of the same translator, who changed some of his usage gradually. We should note that the translator of the *Data* used not only *quod* but also *quoniam* for ὅτι and the translator of the *Elements* used not only *scribere* but also *describere* for γράφειν and the translator of the *Almagest* also used *coniungere* for ἐπιζευγνύναι, though only once. If the same person had gradually changed some translation usage, the order of translation of these works would be as follows:

Data, Optics, Catoptrics → *Elementatio physica* →
Almagest → *Elements*

I agree with Murdoch in his conjecture that the *Elements* would have been translated after the *Almagest*,[65] and indeed the preface to the translation of the *Almagest* suggests that the *Data, Optics, Catoptrics*, and *Elementatio physica* were rendered into Latin before the *Almagest*.

Therefore, whether the *Data* was translated by the same as the translator of the *Almagest* or not, it was probably done before 1160, by which time the *Almagest* is judged to have been translated. The Latin transla-

64) J. Murdoch, *Op. cit.*, pp. 266–267.
65) J. Murdoch, *Op. cit.*, pp. 269–270.

tion of the *Almagest* is assumed to have come out some time close to 1160, because the mission from the Byzantine emperor mentioned in the preface to the translation of the *Almagest* must be none other than the one fol-following the peace of 1158 and because Aristippus was made Archdeacon of Catania in 1156 and died in prison in 1162 after his advancement to a high post in 1160.[66)]

What can we say, then, about the translator of the *Data*? Little is known about him, even if we suppose that he is the same person as the translator of the *Almagest.*

From his preface to the translation of the *Almagest*, we know that he was a stranger to Sicily and that he first studied medicine in Salerno. Further, he tells us that he sought out Aristippus in Sicily to obtain a copy of the *Almagest* which had been brought from Constantinople and that after preliminary study of Euclid's *Data, Optics, Catoptrics*, and Proclus' *Elementatio physica*, he received help from Eugene in his translation of the *Almagest.* In this connection I think it probable that he obtained from Aristippus not only a copy of the *Almagest* but also of Euclid's *Data, Optica*, and *Catoptrics*, for it is known from Aristippus' preface to his translation of Plato's *Phaedo* that he had a copy of the *Optics*.[67)]

66) Haskins, *Op. cit.*, p. 160.

67) "Enricus Aristippus Catinensis archidiaconus Roborato fortune salutem dicit. . . . Habes Euclidis optica, qui tam vere et mirabiliter de opinione videndi disserit, ut opinabilia retiocinatione probet demonstrativa."

Cf. Valentin Rose "Die Lücke im Diogenes Laërtius und der Alte Übersetzer", *Hermes*, Vol. I (1866), p. 387–8.

Rose commented as follows: "Aber noch viele andere litteralish merkwürdige Dinge erfahren wir aus dem Phaedoprolog, z.T. freilich solche die sich der Erläuterung vorläufig entziehen. . . . 'Hier ist Theorides' heißt es, 'hier tuus Aristippus, zu Gebote stehen dir Herons *mechanica*, Euclidis *optica*, die *apodictice* des Aristoteles." (p. 380) "Und auch die unzählige Male im Mittelalter abgeschriebene griechisch lateinische Übersetzung der Optik und Catoptrik des Euclides (*liber de visu* und *de speculis*), hat sie also ihren Ursprung schon im Anfang des 12 Jh. und in Unteritalien?" (p. 381).

Heiberg remarked about that, "Der ganze Zusammenhang macht es wahrscheinlich, daß auch die Hssn.,wonach unser Mann Euklid und Proklos übersetzte, in seinem (Aristippus) Besitz waren; dazu stimmt, daß er in der Vorrede zur Phaidonübersetzung seinem englischen Freund unter den schönen Sachen, die dieser auf Sicilien haben könnte, auch Euklidis optica vorhält. Nun gibt es lateinische Übersetzungen (nach dem Griechischen) von eben den gennanten drei Werken Euklids; man hat sie längst mit Suditalien-Sicilien in Verbindung gebracht; nach dem jetzt Vorliegenden ist es wohl nicht zu kühn, diese Übersetzungen eben unserem Übersetzer, also dem 12. Jahrh. zuzuschreiben."

See J. L. Heiberg "Noch einmal die mittelalterliche Ptolemaios-Übersetzung," *Hermes*, Vol. XLVI, p. 209.

In brief, my present conclusions concerning the identity of the translator are as follows: 1) it is reasonably certain that the author of the Latin translation of Euclid's *Data* also translated the *Optics, Catoptrics*, and Proclus' *Elementatio physica* (or *De motu*); and 2) it is still possible that he is the same person as the anonymous translator of Ptolemy's *Almagest* and Euclid's *Elements.*

As regards the first conclusion, the result of my study agrees well with that of Murdoch, obtained independently. I am confident of it, for it is drawn from full examination of manuscripts and text. As far as the second conclusion is concerned, it is much less well founded. It is based partially on the studies of Haskins and Boese, which identified the translator of the *Almagest* with that of Proclus' work, and also of my textual intuition that the translator of the *Elements* seems to be not different from that of the *Data.* Against the second conclusion, however, there are the cogent arguments presented by Murdoch. We should take them into full consideration. Final determination of this problem should be left for further investigation. It is desirable that the whole text of the Latin *Elements* be published and compared with the whole text of the *Data.* If my edition of the *Data* is useful for this purpose, I would be much gratified.

The above conclusions have considerable importance for clarifying the history of scientific transmission in 12th-century Sicily. If the same man should have translated these various works, he would have occupied a position in editing translations of mathematical works from the Greek not unlike that of Gerard of Cremona in the sphere of translations from the Arabic.

(iii) Comparison among the Manuscripts of Oxford, Paris, Dresden and Berlin

As I have already mentioned, the four manuscripts from which I edited the text of the medieval Latin translation of Euclid's *Data* are the following:

1) MS Oxford, Bodl. Auct. F.5.28, 13c, 99r-113r (cat. pag. 58r-72r)
2) MS Paris, Bibl. Nat. lat. 16648, 13–14c, 60r-91v
3) MS Dresden, Sächs. Landesbibl. Db 86, 14c, 200r-213r
4) MS Berlin. Staatsbibl. Preuß. Kulturbesitz, Q 510 14c, 178v-191r

For convenience, I will refer to MS Oxford as MS *O*, MS Paris as MS *P*, MS Dresden as MS *D*, and MS Berlin as MS *B*.

First of all, I should say that there is no significant difference in content

among these manuscripts. They are copies of the same Latin translation of the *Data*.

MS *O* is the best of these four manuscripts. It is copied so carefully that I found few omissions. It is also accurately written except for a few scribal errors. In MS *O*, the same hand as that of the text has inserted numerous interlineal and marginal glosses in the form of references to definitions and preceding propositions in the *Data* and also sometimes to propositions in the *Elements*. There are also some deletions by the same hand. It seems probable that the scribe checked over his copy after completing it. It is of interest that MS *O* includes a marginal preface and long notes on the proofs of Propositions 26, 39, and 67, in the same hand as that of the text. MS *O* includes also in a later hand the title "*Liber de datis magnitudinibus*" as well as some glosses, almost all of which are too faded to be legible. The principal hand is Gothic of the middle 13c, and fairly extensive abbreviation is employed. The text is written in a single column. The diagrams in the margins are accurately drawn in the horizontal direction.

A fairly good manuscript in comparison with MS *D* and *B*, MS *P* is much more corrupt than MS *O*; one often cannot understand the readings in MS *P* without reference to MS *O*. One also finds many omissions and scribal errors there which MS *O* preserves, though at the same time, there are a few errors in MS *O* which are not found in MS *P*. It contains neither title nor interlineal glosses. MS *P* is written in a single column in a clear Gothic hand with comparatively few abbreviations. The figures in the margins are somewhat imprecisely drawn and oriented vertically.

MS *D*, which is also written in handsome Gothic, is so terribly damaged (probably by the wartime bombing of the city of Dresden) that one cannot read even half of the text. It has many marginal glosses written in the same hand as that of the text, just as in MS *O*. It contains many reiterations and scribal errors. There are no numbering of definitions and propositions in MS *D*. The diagrams drawn horizontally are also so damaged that almost all parts of them have faded away.

We may say that MS *B*, written in acute Gothic, is the most corrupt among the four manuscripts. It has many more omissions and scribal errors than any of the others. There is no numbering of definitions in MS *B*, and the last part of it is replaced by the text of some other work. It has diagrams only up to proposition 15; later no figures are drawn at all.

The relation among these four manuscripts is really a difficult problem to solve. But at least we can draw the following conclusions: MS *B* often has common omissions with MS *P*, though the former contains many

more omissions and scribal errors than the latter. Thus MS *B* seems to be more akin to MS *P* than to the other manuscripts. MS *D* has some glosses identical with those in MS *O*, though the former lacks many other glosses contained in the latter. It is also more corrupt than MS *O* in its reiterations and omissions. It should also be noted that there are a number of errors common to all the manuscripts which seem to have a common origin in an earlier manuscript. Consequently, a possible geneology of these manuscripts could be like this:

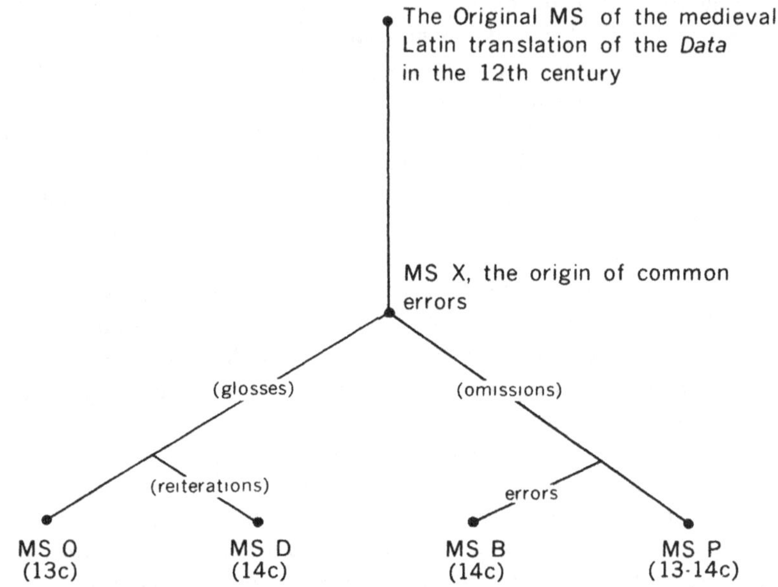

I have inserted specimen plates of the four manuscripts at the end of this section.

Lastly I should mention the Greek manuscript of the *Data* on which our medieval Latin translation was based.

I think that it was translated from a Greek manuscript of the text as it existed before Theon's revision. This I conclude from the fact that our Latin translation accords very closely with Menge's Greek text which was edited mainly on the basis of the old non-Theonine manuscript discovered by Peyrard, namely, MS Vat. Gr. 190. But it should be noted, at the same time, that there is not so great difference between Theonine and non-Theonine texts of the *Data* as those of the *Elements* and also that our Latin text happens to coincide in a few cases with Menge's Theonine texts As regards comparison with Greek manuscripts, I note peculiar agree-

ments under the translation as far as I can notice them. In this connection I also add the fact that our Latin translation does not contain the *vulgo propositio* LXXX VII and the subsequent lemma in the Vatican manuscript of Peyrard, which is contained in the appendix of Menge's text.

In the following, I give some examples of Menge's Greek text of the *Data* and our medieval Latin translation in the parallel arrangement. The close relation between them will be very obvious.

DATA

Ἐὰν δύο μεγέθη πρὸς ἄλληλα λόγον ἔχῃ δεδομένον καὶ ἀφαιρεθῇ ἀπὸ ἑκατέρου αὐτῶν δεδομένον μέγεθος, τὰ λοιπὰ πρὸς ἄλληλα ἤτοι λόγον ἕξει δεδομένον, ἢ τὸ ἕτερον τοῦ ἑτέρου δοθέντι μεῖζόν ἐστιν ἢ ἐν λόγῳ

Δύο γὰρ μεγέθη τὰ AB, ΓΔ πρὸς ἄλληλα λόγον ἐχέτω δεδομένον, καὶ ἀφῃρήσθω ἀφ' ἑκατέρου αὐτῶν δεδομένον μέγεθος, ἀπὸ μὲν τοῦ AB τὸ EA, ἀπὸ δὲ τοῦ ΓΔ τὸ ΓZ· λέγω, ὅτι τὰ λοιπὰ τὰ EB, ZΔ πρὸς ἄλληλα ἤτοι λόγον ἕξει δεδομένον, ἢ τὸ ἕτερον τοῦ ἑτέρου δοθέντι μεῖξόν ἐστιν ἢ ἐν λόγῳ.

Ἐπεὶ γὰρ ἑκάτερον τῶν AE, ΓZ δοθέν ἐστι, λόγος ἄρα τοῦ AE πρὸς ΓZ δοθείς, καὶ εἰ μὲν ὁ αὐτός ἐστι τῷ τοῦ AB πρὸς ΓΔ, ἔσται καὶ λοιποῦ τοῦ EB πρὸς λοιπὸν τὸ ZΔ λόγος δοθείς.

Μὴ ἔστω δὴ ὁ αὐτός, καὶ πεποιήσθω ὡς τὸ AB πρὸς ΓΔ, οὕτως τὸ AH πρὸς ΓZ. λόγος δὲ τοῦ AB πρὸς τὸ ΓΔ δοθείς· λόγος ἄρα καὶ τοῦ AH πρὸς τὸ ΓZ δοθείς· δοθὲν δὲ τὸ ΓZ· δοθὲν ἄρα καὶ τὸ AH. ἔστι δὲ καὶ τὸ AE δοθέν· καὶ λοιπὸν ἄρα τὸ EH δοθέν ἐστιν. καὶ ἐπεὶ ὡς τὸ AB πρὸς τὸ ΓΔ, οὕτως τὸ AH πρὸς τὸ ΓZ, λοιποῦ ἄρα τοῦ HB πρὸς λοιπὸν τὸ ZΔ λόγος ἐστὶ δοθείς. καί ἐστι δοθὲν τὸ EH· τὸ EB ἄρα τοῦ ZΔ δοθέντι μεῖζόν ἐστιν ἢ ἐν λόγῳ.

XV. Si due quantitates adinvicem proportionem habuerint datam, auferaturque ab utraque earum data quantitas, reliqua adinvicem proportionem habebunt datam, vel alterum altero dato maius est quam in proportione.

Due enim quantitates AB et GD adinvicem proportionem habeant datam, auferaturque ab utraque earum data quantitas, ab AB EA et ab GD GZ. Dico quod reliqua EB et ZD adinvicem vel proportionem habebunt datam vel alterum altero dato maius est quam in proportione.

Quoniam enim utrumque AE, GZ datum est, proportio ergo AE ad GZ data. Et si eadem est quantitatis AB ad GD, erit et reliqui EB ad reliquum ZD proportio data.

Non sit autem eadem, fiatque sicut AB ad GD ita AI ad GZ. Atqui proportio quantitatis AB ad GD data. Proportio ergo et quantitatis AI ad GZ data, datumque GZ. Datum ergo et AI. Estque et AE datum. Et reliquum ergo EI datum est. Et quoniam sicut AB ad GD ita AI ad GZ, reliqui ergo IB ad reliquum ZD proportio data. Et est datum EI. Quantitas ergo EB quantitate ZD dato maior est quam in proportione.

PLATES

The following plates are photographic copies of two folios from each of the four manuscripts of the medieval Latin translation of Euclid's *Data*, which have been used in the preparation of our text in Part II.

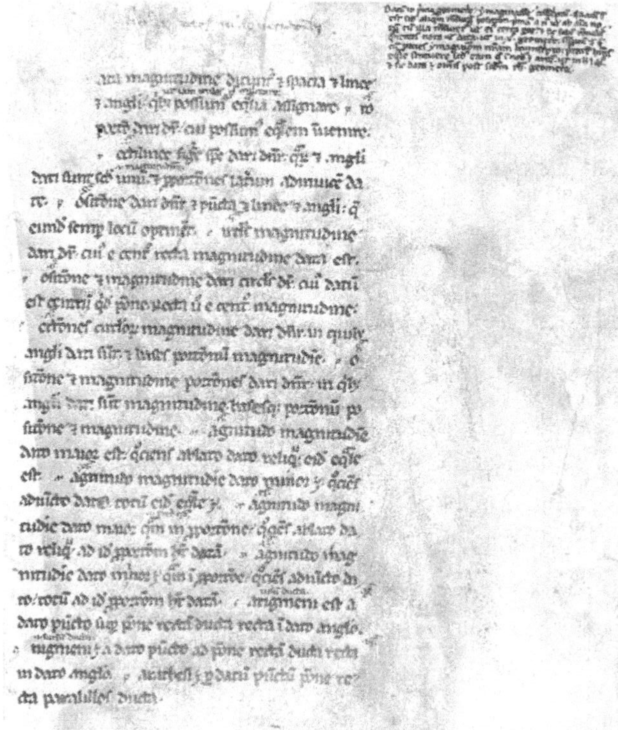

99r (cat. pag. 58r)

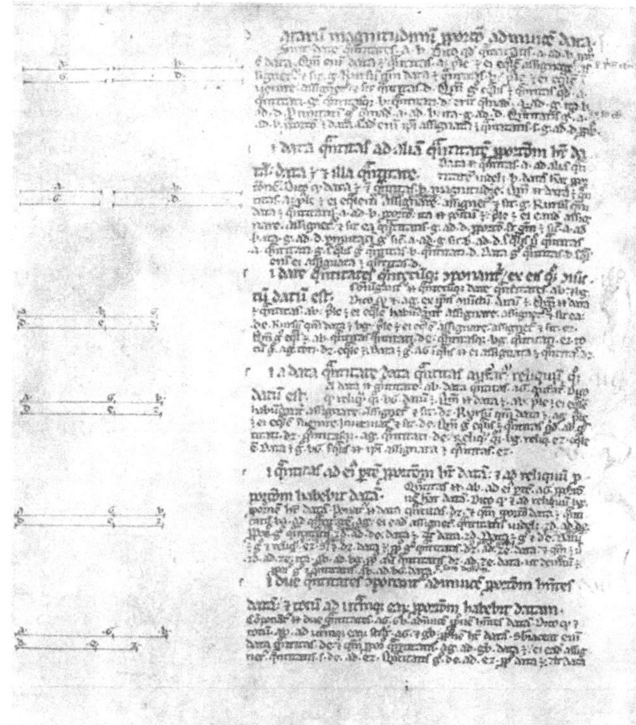

99v (cat. pag. 58v)

MS Oxford, Bodl. Auct. F.5.28, 13c.

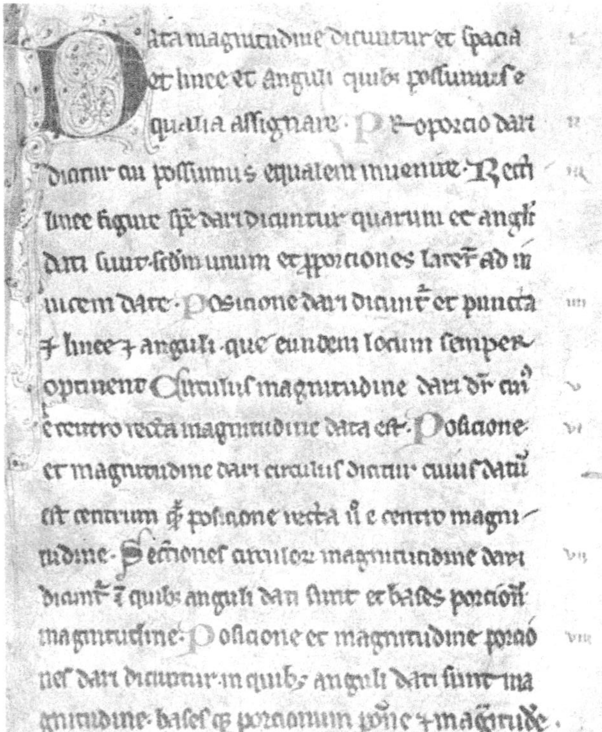

60r

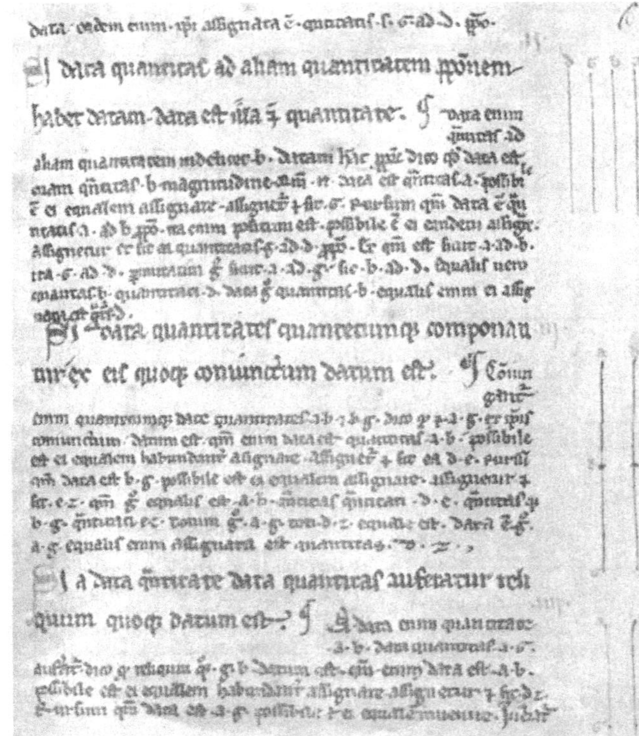

61r

MS Paris, Bibl. Nat. lat. 16648, 13–14c.

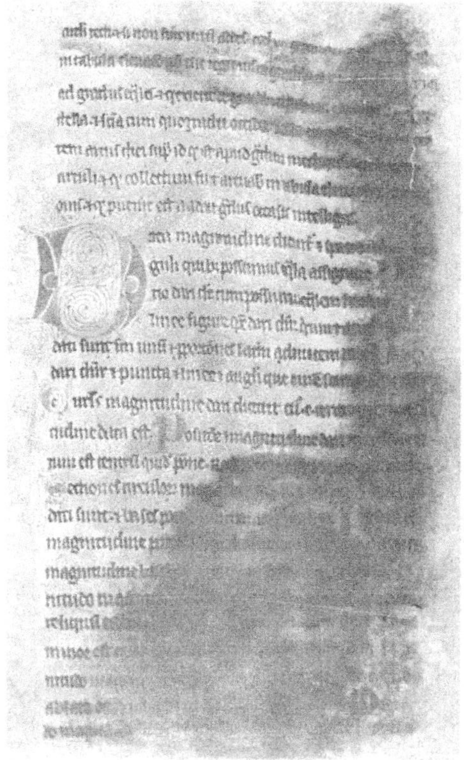

200r

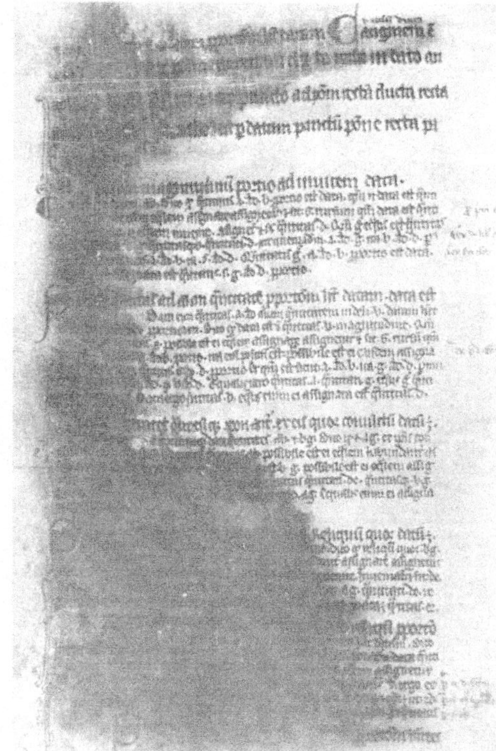

200v

MS Dresden, Sächs. Landesbibl. Db 86, 14c.

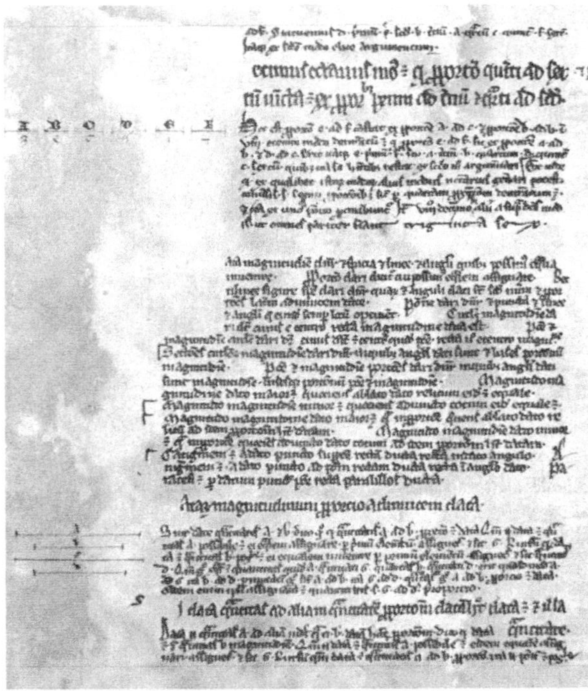

178v

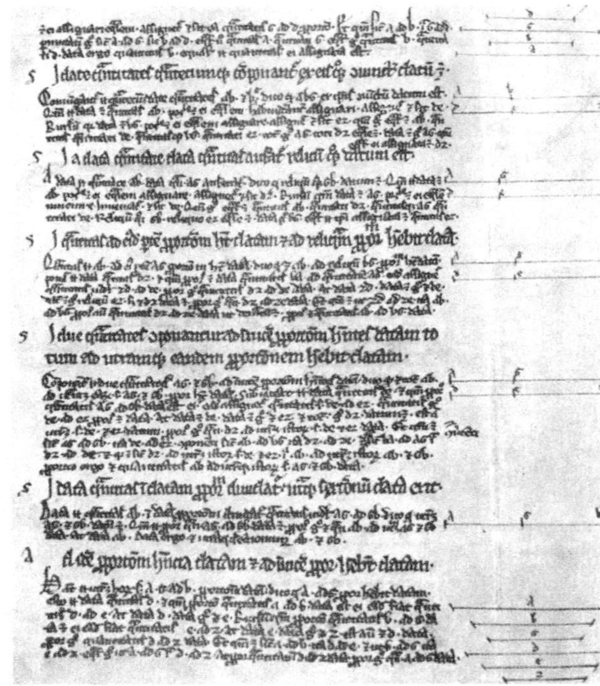

179r

MS Berlin, Staatsbibl. Preuß. Kulturbesitz Q 510, 14c.

PART II

TEXT OF THE MEDIEVAL LATIN VERSION OF THE *DATA* AND ITS ENGLISH TRANSLATION

CHAPTER 4

GENERAL REMARKS ON THE EDITION AND TRANSLATION

Before presenting the Latin text of the *Data* and its English translation, it seems pertinent to make some general remarks on the procedures and techniques of editing, collating, and translating which I have employed.

Since, as I remarked in the last section of Chapter 3, the Oxford manuscript is the best and most accurate in content, I took it as the basis for my edition and used the other three manuscripts principally as aids in reading this basic text. Therefore folio numbers in margin of the text are those (cat pag.) of the Oxford manuscript. However, the four manuscripts were fully collated to obtain the best readings.

The capitalization, punctuation, and paragraphing are my own. According to the Euclidean manner, I have paragraphed the text in the order of 1) "enunciation" (*πρότασις*), 2) "example" (*ἔκθεσις*) and "the specification of the proposition" (*διορισμός*), 3) "further construction" (*κατασκευή*) and "demonstration" (*ἀπόδειξις*), 4) other demonstrations. The enunciation of each proposition has been expressed in larger letters to set it off from the proof. I also capitalized the letters standing for geometrical magnitudes and points to avoid confusion, although they are written in small letters in the manuscripts. For convenience, the Roman numerals numbering propositions, which in the Oxford manuscript are written in the right hand margin next to the enunciations, have been transposed in my text before the enunciations.

All of the variant readings, glosses, and omissions have been exactly recorded, including orthographical diversions. But the glosses in the second hand in the Oxford manuscript, which are few in number, have been ignored, because they are illegible. Damaged parts of the text in the Dresden manuscript are noted as illegible (= *illeg.*). As regards glosses in the same manuscript, which were saved from damage, I note them in collation when the parts of the text they refer to are legible, but not when the text they refer to is destroyed, because it is not clear what parts they refer to.

Angle brackets ⟨ ⟩ represent additions to the text to restore what I believe to be original readings. This restoration is done in comparison with the Greek original.

In my English translation of the edited text I have kept as close as possible to the Latin sentence structure within the limits of natural English. However, I have not always retained the grammatical correspondences in mood and tense, because, for example, the subjunctive mood in Latin is not always the same as that in English.

Moreover, I have often used mathematical notations to express mathematical relations briefly. For example, I have written "A:B = C:D" in place of the rhetorical statement "the ratio of A to B is equal to the ratio of C to D".

I have inserted diagrams at the side of the translation in the horizontal direction in the manner of the Oxford manuscript, for the figures in that manuscript are drawn more accurately and clearly than those in the other manuscripts. Still I have sometimes modified them so that they accord with the content of the text.

Square brackets [ ] indicate my own supplements in the translation to make the meaning clearer.

For this translation, I have consulted the earlier English versions of the *Data* by Simson and Haselden.

Major Latin Terms and Abbreviations Used in the Variant Readings

add. (addidit)	has added
ante	before
bis	twice
cf. (confer)	confer
corr. ex (correxit ex)	has corrected from
correxi ex	I have corrected from
del. (delevit)	has deleted
delevi	I have deleted
hic et ubique	here and everywhere
illeg. (illegibilis)	illegible
ln. (*in linea*)	in the line
mg. (in margine)	in the margin
manu rec. (manu recentiore)	in a later hand
om. (omisit)	has omitted
post	after
scr. (scripsit)	has written
sed	but
sic delendum	is to be so deleted
supra	above
tr. (transposuit)	has transposed

Sigla of Manuscripts

O—MS Oxford, Bodl. Auct. F.5.28, 99r-113r, 13c.
P—MS Paris, Bibl. Nat. lat. 16648, 60r-91v, 13–14c.
D—MS Dresden, Sächs. Landesbibl. Db 86, 200r-213r, 14c.
B—MS Berlin, Staatbibl. Preuß. Kulturbesitz, Q 510, 178v-191r, 14c.

CHAPTER 5

THE EDITED LATIN TEXT OF THE *DATA* AND ITS ENGLISH TRANSLATION

<DIFFINITIONES>

(58r) I./Data magnitudine dicuntur et spatia et linee et anguli, quibus possumus equalia assignare.

II. Proportio dari dicitur, cui possumus equalem invenire.

III. Rectilinee figure specie dari dicuntur, quarum et anguli dati sunt secundum unum et proportiones laterum adinvicem date.

IIII. Positione dari dicuntur et puncta et linee et anguli, que eundem semper locum optinent.

V. Circulus magnitudine dari dicitur, cuius e centro recta magnitudine data est.

VI. Positione et magnitudine circulus dari dicitur, cuius datum est centrum quidem positione, recta vero e centro magnitudine.

VII. Sectiones circulorum magnitudine dari dicuntur, in quibus anguli dati sunt et bases portionum magnitudine.

1 *in mg. super. habet O manu rec. titulum* Liber de datis magnitudinibus / *in angulo super. add. O* Datum in prima geometrie ymaginatione circumscriptum. Datum etiam est secundum aliquam mensuram positam primam a natura vel ab anima notum cum illa mensuretur vel eius certa parte et sic solum rationalis quantitas nota vel data, ut in X° geometrie. Datum etiam est cum penes ymaginationem nostram circumscripto: potestatem habemus equale statuere secundum eam quod in nobis est artem ut hoc libro; et sic data est omnis post secundum thematem geometrie.

2 I *om. BDO;* numeros diffinitionum *scr. mg. P et om. BD hic et ubique* / spat[ia....an]guli *illeg. D* / spatia *correxi ex* spacia *BOP hic et ubique*

2–3 *supra* quibus possumus equalia *add. O* ut iam miles potest militare

3 equalia: equala *D* / assignare: invenire *B*

3–4 assign[are....Propor]tio *illeg. D*

4 II: 2ª *O*; numeros diffinitionum *supra scr. O hic et ubique* / Proportio: Proporcio *P hic et ubique* / cui: cum *D* / invenire: assignare *B*

4–5 invenire....Rectil]linee *illeg. D*

5 III: 3ª *O* / et anguli *illeg. D*

6 *supra* unum *add. O* magnitudinem / adinvicem: ad invicem *P hic et ubique*

6–7 date....Positione *illeg. D*

7 IIII: 4ª *O* / Positione: Posicione *P hic et ubique*

8 s[emper....optinent *illeg. D*

9 V: 5ª *O* / cen[tro....magni]tudine *illeg. D*

11 VI: 6ª *O* / magni[tudine....da]tum *illeg. D* / circulus *tr. OP post* dari

12 recta....magnitudine *illeg. D*

13 VII: 7ª *O* / m[agnitudine....anguli *illeg. D*

14 portionum: porcionum *P hic et ubique*

14–1(p.s.) por[tionum....et *illeg. D*

DEFINITIONS

I. Spaces, lines and angles are said to be given in magnitude, when we can assign equals to them.

II. A ratio is said to be given, when we can find an equal to it.

III. Rectilineal figures are said to be given in species, when each of their angles is given and also the ratios of their sides to one another are given.

IV. Points, lines and angles are said to be given in position, when they always hold the same place.

V. A circle is said to be given in magnitude, when its straight line from the center [*i.e.*, its radius] is given in magnitude.

VI. A circle is said to be given in position and magnitude, when its center is given in position and its straight line from the center [*i.e.*, its radius] is given in magnitude.

VII. Segments of circles are said to be given in magnitude, when the angles in them and the bases of the segments are given in magnitude.

VIII. Positione et magnitudine portiones dari dicuntur, in quibus anguli dati sunt magnitudine basesque portionum positione et magnitudine.

IX. Magnitudo magnitudine dato maior est, quotiens, ablato dato, reliquum eidem equale est.

X. Magnitudo magnitudine dato minor est, quotiens, adiuncto dato, totum eidem equale est.

XI. Magnitudo magnitudine dato maior est quam in proportione, quotiens, ablato dato, reliquum ad idem proportionem habet datam.

XII. Magnitudo magnitudine dato minor est quam in proportione, quotiens, adiuncto dato, totum ad idem proportionem habet datam.

XIII. Catigmeni est a dato puncto super positione rectam ducta recta in dato angulo.

XIIII. Anigmeni est a dato puncto ad positione rectam ducta recta in dato angulo.

XV. Parathesi est per datum punctum positione recta parallelos ducta.

1 VIII: 8[a] *O*

1–2 po[rtiones. . . . sunt *illeg. D*

2–3 basesque. . . . Mag]nitudo *illeg. D*

3 IX: 9[a] *O* / m[agnitudine. . . . dato[2] *illeg. D* / quotiens *correxi ex* quociens *BOP hic et ubique* /

4 reliquum: relicuum *B* / equale *tr. B post* est

4–5 reliquum. . . . dato[1] *om. P* / eidem . . . . dato[1] *illeg. D*

5 X: XI[a] *O* / dato[1] *om. B* / dato[2] *om. B*

5–7 est. . . . Mag]nitudo *illeg. D*

7 XI *correxi ex* 12[a] *O* X *P* / *post* magnitudine *scr. P* dato minor est quam in proportione quociens adiuncto / est *om. O* / *supra* quam in *add. O* preter quod

7–8 magnitudine. . . . quotiens *illeg. D*

8 reliquum: relicuum *B*

8–9 abl[ato. . . . Magnitu]do *illeg. D*

9 XII *correxi ex* 13[a] *O* XI *P*

9–10 ma[gnitudine. . . . datam *illeg. D*

11 XIII *correxi ex* 14[a] *O* XII *P* / *numerum scr. mg. O* / *supra* Catigmeni *add. O* ursum ducta *et D* rursum ducta / a. . . . recta *illeg. D* / rectam: recta *P*

12–13 an[gulo. . . . puncto *illeg. D*

13 XIII *correxi ex* 15[a] *O* XIII *P* / *numerum scr. mg. O* / *supra* Anigmeni *add. O* sursum ducta / positione: positionem *BD*

13–15 in. . . . est *illeg. D*

14 dato *tr. B post* angulo

15 XV *correxi ex* XIIII *P* / Parathesi: Paratesi *B* / parallelos *correxi ex* paralillos *BOP* / pa[rallilos ducta *illeg. D*

VIII. Segments [of circles] are said to be given in position and in magnitude, when the angles in them are given in magnitude and the bases of the segments are given in position and magnitude.

IX. A magnitude is greater than another magnitude by a given [magnitude], when, this given [magnitude] being taken from it, the remainder is equal to the other magnitude.

X. A magnitude is less than another magnitude by a given [magnitude], when, this given [magnitude] having been added to it, the whole is equal to the other magnitude.

XI. A magnitude is greater than another magnitude by a given [magnitude] as in ratio,* when, the given [magnitude] having been subracted, the remainder has a given ratio to the same [magnitude].

XII. A magnitude is less than another magnitude by a given [magnitude] as in ratio,* when, the given [magnitude] having been added to it, the whole has a given ratio to the same [magnitude].

XIII. A straight line is [said to be] catigmeni [*i.e.*, lowered], when it is drawn down from a given point on a straight line [given] in position, making a given angle.

XIV. A straight line is [said to be] anigmeni [*i.e.*, elevated], when it is drawn up from a given point to a straight line [given] in a position, making a given angle.

XV. A straight line is [said to be] parathesi [*i.e.*, juxtaposed], when it is drawn, through a given point, parallel to a straight line [given] in position.

* For the peculiar notion "to be greater or less as in ratio", cf. note 1 on page 227. This notion appears often from Proposition X on.

<PROPOSITIONES>

(58v) /I. Datarum magnitudinum proportio adinvicem data.

Sint date quantitates *A*, *B*. Dico quod quantitatis *A* ad *B* proportio est data.

Quoniam enim data est quantitas *A*, possibile est ei equalem assignare. Assignetur et sit *G*. Rursum quoniam data est quantitas *B*, possibile est ei equalem invenire. Assignetur et sit quantitas *D*. Quoniam ergo equalis est quantitas quidem *A* quantitati *G*, quantitasque *B* quantitati *D*, erit quemadmodum *A* ad *G* ita *B* ad *D*. Permutatim ergo quemadmodum *A* ad *B* ita *G* ad *D*. Quantitatis ergo *A* ad *B* proportio est data, eadem enim ipsi assignata est quantitatis scilicet *G* ad *D* proportio.

II. Si data quantitas ad aliam quantitatem proportionem habet datam, data est etiam illa quantitate.

Data enim quantitas *A* ad aliam quantitatem videlicet *B* datam habeat proportionem. Dico quod data est etiam quantitas *B* magnitudine.

Quoniam enim data est quantitas *A*, possibile est ei equalem assignare. Assignetur et sit *G*. Rursum quoniam data est quantitatis *A* ad *B* propor-

2 I: XV *P om. BD; numeros propositionum scr. mg. OP et om. D hic et ubique* / Datarum: Patarum *P illeg. D* / proportio: proporcio *B*
3 *ante* Sint *scr. P* I / Sint.... quantitates *illeg. D* / *post* A *scr. B* et / *post* Dico *scr. B* ergo
5 quanti[tas....equa]lem *illeg. D* / *post* assignare *scr. B* per primum elementum *et add. mg. O* per primum elementum huius
6 quoniam: quia *B*
6–7 quanti[tas....equale]m *illeg. D*
7 *post* invenire *scr. B* per primum elementum
8 quidem....quan]titasque *illeg. D* quantitasque: quantitas *B*
9 Per[mutatim....quemadmodum *illeg. D* / *de* Per- *add. mg. D* ex V libro elementorum / queadmodum[2]: sicut *B*
10 *de* Quantitatis *add. mg. D* ex secundo elementorum / *post* ergo *add. mg. O* ex secundo elemento
10–11 eadem....assigna]ta *illeg. D*
12 II om. *B* / Si . . . . quanti]tas *illeg. D*/habet *tr. B post* datam *ln.13*
13 etiam: et *BO* / etiam *tr. P post illa* / eitam. . . . quantitate *illeg. D*
14 A *om. P* / quantitatem *tr. B post* videlicet
16 enim....equantitas *illeg. D* / ei: eidem *B* / assignare: assignari *B* / *supra* assignare *add. O* ex prima diffiinitione
17 data....A *illeg. D*

PROPOSITIONS

I. The ratio of given magnitudes to one another is given.

Let A and B be given quantities. I say that the ratio $A:B$ is given.

Since the quantity A is given, it is possible to assign an equal to it. Let this be assigned and be G. Further, since the quantity B is given, it is possible to find an equal to it. Let it be assigned and be the quantity D. Since the quantity A is equal to the quantity G and the quantity B to the quantity D, then $A:G = B:D$. By permutation, $A:B = G:D$. Therefore, the ratio $A:B$ is given, for a ratio equal to it has been assigned, namely, the ratio $G:D$.

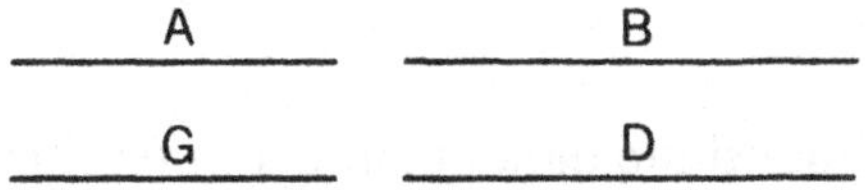

II. If a given quantity has a given ratio to another quantity, the latter is also given in quantity.

Let the given quantity A have this given ratio to another quantity, namely, B. I say that the quantity B is also given in magnitude.

Since the quantity A is given, it is possible to assign an equal to it. Let it be assigned and be G. Further, since the ratio $A:B$ is given, for it has been

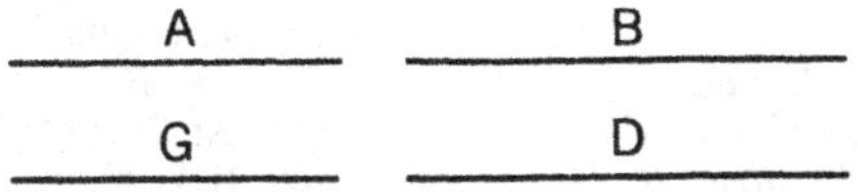

tio, ita enim positum est, possibile est ei eandem assignare. Assignetur et sit ea quantitatis *G* ad *D* proportio. Et quoniam est sicut *A* ad *B* ita *G* ad *D*, permutatim ergo sicut *A* ad *G* sit *B* ad *D*. Equalis vero quantitas *A* quantitati *G*. Equalis ergo quantitas *B* quantitati *D*. Data ergo quantitas *B*, equalis enim ei assignata est quantitas *D*.

III. Si date quantitates quantecumque componantur, ex eis quoque coniunctum datum est.

Coniungantur enim quantecumque date quantitates *AB* et *BG*. Dico quod et *AG* ex ipsis coniunctum datum est.

Quoniam enim data est quantitas *AB*, possibile est ei equalem habundanter assignare. Assignetur et sit ea *DE*. Rursum quoniam data est *BG*, possibile est ei equalem assignare. Assignetur et sit *EZ*. Quoniam ergo equalis est *AB* quantitas quantitati *DE* quantitasque *BG* quantitati *EZ*, totum ergo *AG* toti *DZ* equale est. Data est ergo *AG*, equalis enim ei assignata est quantitas *DZ*.

IIII. Si a data quantitate data quantitas auferatur, reliquum quoque data est.

A data enim quantitate *AB* data quantitas *AG* auferatur. Dico quod reliquum quoque *GB* datum est.

Quoniam enim data est *AB*, possibile est ei equalem habundanter assi-

1 *de* possibile *add. mg. D* ex premissa diffinitione / eandem assignare: assignari equalem *B*
1–2 assigna[re. . . .quantita]tis *illeg. D*
2 est *om. B*
3 permu[tatim. . . .G *illeg. D*
3–4 A[2]. . . .quantitas[1] *om. P*
4 quanti[tas[1]. . . .D *illeg. D*
5 quantitas *tr. B ante* ei / D *om. B*
6 III: 3^{a} *B* / Si. . . . quantita]tes *illeg. D*
8 Coniungantur. . . .quantecumque *illeg. D*
9 AG:ABG *B*
9–10 con[iunctum. . . .AB *illeg. D*
11 assignare: assignari *B* / as[signare. . . .e]st *illeg. D* / quoniam: quia *B*
12–13 assig[nare. . . .quan]titas *illeg. D*
13–14 quantitati[2]. . . .ergo[2] *illeg. D*
14 *ante* equalis *scr. B* quoniam / enim *om. B*
15 est. . . .DZ *illeg. D* / quantitas *om. B*
16 IIII: 4^{a} *B* / Si. . . . auferatur *illeg. D* / reliquum: relicuum *B*
18 A. . . .auferatur *illeg. D*
19 reliquum: reliqum *B hic et ubique* / GB: BG *DO*
19–20 datum. . . .habundanter *illeg. D*
20 habundanter *om. B*

assumed so, it is possible to assign a ratio equal to it. Let it be assigned and be the ratio $G:D$. And since $A:B = G:D$, by permutation, $A:G = B:D$. But indeed the quantity A is equal to the quantity G. Therefore, the quantity B is equal to the quantity D. Therefore, the quantity B is given, for the quantity D, equal to it, has been assigned.

III. If any given quantities are added together, the quantity composed of them is also given.

Let any given quantities *AB* and *BG* be conjoined. I say that the quantity *AG* composed of them is also given.

Since the quantity *AB* is given, it is possible to assign an equal to it. Let this be assigned and be *DE*. Further, since *BG* is given, it is possible to assign an equal to it. Let this be assigned and be *EZ*. Since the quantity *AB* is equal to the quantity *DE* and the quantity *BG* is equal to the quantity *EZ*, hence the whole *AG* is equal to the whole *DZ*. Therefore, *AG* is given, for the quantity *DZ* equal to it has been assigned.

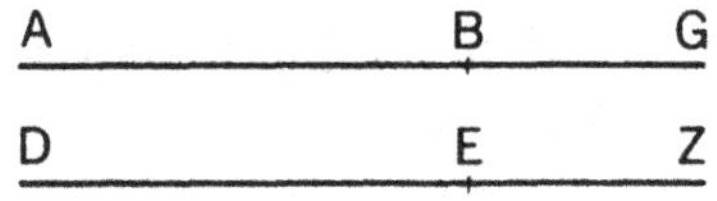

IV. If a given quantity is taken from a given quantity, the remainder is also given.

From the given quantity *AB*, let the given quantity *AG* be taken away. I say that the remainder *GB* is also given.

Since *AB* is given, it is possible to assign an equal to it. Let this be

gnare. Assignetur et sit *DZ*. Rursum quoniam data est *AG*, possibile est ei equalem invenire. Inveniatur et sit *DE*. Quoniam ergo equalis est quantitas quidem *AB* quantitati *DZ* quantitasque *AG* quantitati *DE*, reliquum quoque *BG* reliquo *EZ* equale est. Data est ergo *BG*, equalis enim ipsi assignata est quantitas *EZ*.

V. Si quantitas ad eius partem proportionem habet datam, et ad reliquum proportionem habebit datam.

Quantitas enim *AB* ad eius partem *AG* proportionem habeat datam. Dico quod et ad reliquum *BG* proportionem habet datam.

Ponatur enim data quantitas *DZ*. Et quoniam proportio data est quantitatis *BA* ad quantitatem *AG*, ei eadem assignetur quantitatis videlicet *ZD* ad *DE*. Proportio ergo quantitatis *ZD* ad *DE* data est. At data *ZD*. Data est ergo et *DE*. Datum est ergo et reliquum *EZ*. Sed et *DZ* data est. Proportio ergo quantitatis *DZ* ad *ZE* data. Et quoniam est ut *ZD* ad *ZE* ita *AB* ad *BG*, proportio autem quantitatis *DZ* ad *ZE* data, ut demonstratum est, proportio ergo et quantitatis *AB* ad *BG* data.

VI. Si due quantitates componantur adinvicem proportionem habentes datam, et totum ad utramque earum proportionem habebit datam.

Componantur enim due quantitates *AG*, *GB* adinvicem proportionem

1–2 et....inve]nire *illeg. D*
2–3 Quoniam...quantitasque *illeg. D*
3 quidem *om. B*
4–5 re[liquum....assigna]ta *illeg. D*
6 V: 5^a *B* / eius: eiusdem *B*
6–7 Si. . . . reli]quum *illeg. D*
7 reliquum: reliquam *B*
7–8 habebit. . . . datam *illeg. D*
8 *ante* ad *scr. P* et / habeat: habet *B*
9 *post* et *scr. B* AB
9–10 quod....enim *illeg. D*
10–11 quanti[tas....ass]gnetur *illeg. D*
11–13 quantitatis....est^1 *illeg. D*
12 ZD^1: DZ *B* / *supra* ZD^1 ad *add. O* ex secunda diffinitione
13 *supra* DE *add. O* per II proportionem / et^2 *om. P* / *supra* EZ *add. O* per IIII propositionem
13–14 DE....quoniam *illeg. D*
14 *supra* DZ ad ZE *add. O* per primam proportionem / ut: uti *D*
14–16 ad^2....quanti]tatis *illeg. D*
16 ergo *om. B* / quantita[tis....data *illeg. D* / *post* data *scr. P* est / *post* data *add. O* per secundam diffinitionem
17 VI: 6^a *B* / Si. . . . proporti]onem *illeg. D*
18 et *om. B* / earum: eandem *B*
18–19 propor[tionem...datam *illeg. D*
20 *post* AG *scr. B* et
20–1(p.s.) quantitates....AB *illeg. D*

assigned and be *DZ*. Further, since *AG* is given, it is possible to find an equal to it. Let this be found and be *DE*. Then, since the quantity *AB* is equal to the quantity *DZ* and the quantity *AG* is equal to the quantity *DE*, the remainder *BG* is also equal to the remainder *EZ*. Therefore *BG* is given, for the quantity *EZ* equal to it has been assigned.

V. If a quantity has a given ratio to a part of it, it will also have a given ratio to the remainder.

Let the quantity *AB* have a given ratio to its part *AG*. I say that it [*AB*] also has a given ratio to the remainder *BG*.

Let a quantity *DZ* be given. And, since the ratio *BA*:*AG* is given, let a ratio equal to it be assigned, namely *ZD*:*DE*. Then the ratio *ZD*:*DE* is given. Now *ZD* is given. Therefore DE is also given. Therefore, the remainder *EZ* is also given. But *DZ* is also given. Therefore the ratio *DZ*:*ZE* is given. And since *DZ*:*ZE* = *AB*:*BG* and the ratio *DZ*:*ZE* is given, as has been demonstrated, the ratio *AB*:*BG* is given.

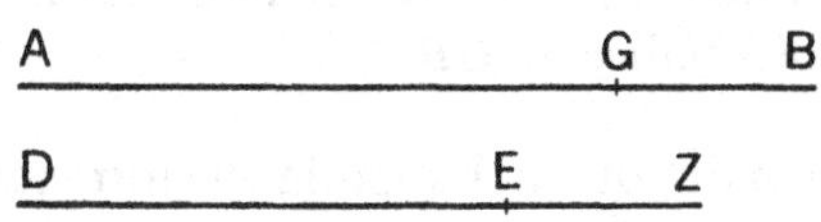

VI. If two quantities which have a given ratio to one another are conjoined, the whole will also have a given ratio to each of them.

Let the two quantities *AG* and *GB*, which have a given ratio to one

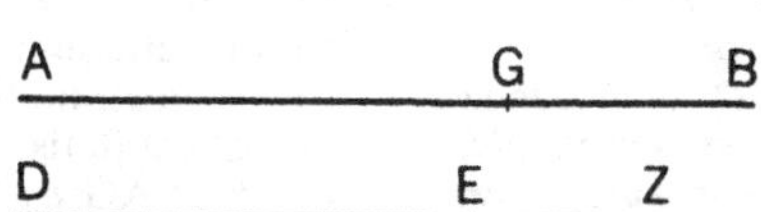

habentes datam. Dico quod et totum *AB* ad utramque earum scilicet *AG* et *GB* proportionem habet datam.

Subiaceat enim data quantitas *DE*. Et quoniam proportio quantitatis *AG* ad *GB* data est, ei eadem assignetur quantitatis scilicet *DE* ad *EZ*. (59r) Quantitatis ergo *DE* ad *EZ* proportio data est. At data/est *DE*. Data ergo et *EZ*. Et totum ergo *DZ* datum est. Est autem utrumque scilicet *DE* et *EZ* datum. Proportio ergo quantitatis *DZ* ad utrumque istorum scilicet *DE* et *EZ* data. Et quoniam est sicut *AG* ad *GB* ita *DE* ad *EZ*, componenti sicut *AB* ad *BG* ita *DZ* ad *ZE*. Et convertenti sicut *BA* ad *AG* ita *DZ* ad *ED*. Et quoniam est sicut *DZ* ad utrumque istorum scilicet *DE* et *EZ* ita *AB* ad utrumque istorum scilicet *AG* et *GB*, proportio ergo et quantitatis *AB* ad utrumque istorum scilicet *AG* et *GB* data.

VII. Si data quantitas in datam proportionem dividatur, utraque sectionum data erit.

Data enim quantitas *AB* in datam proportionem dividatur quantitatis videlicet *AG* ad *GB*. Dico quod utrumque *AG* et *GB* datum est.

Quoniam enim proportio quantitatis *AG* ad *GB* data est, proportio ergo et quantitatis *AB* ad utrumque *AG* et *GB* data. At data *AB*. Data ergo et utraque sectionum *AG* scilicet et *GB*.

VIII. Ad idem proportionem habentia et adinvicem proportionem habebunt datam.

1–3 earum. . . .proportio *illeg. D*
3 quantitatis: quantitatum *P*
4–5 ei. . . .DE[1] *illeg. D*
5 data[1] est: est data *B* / est[2]: etiam *DOP* / *post* Data *scr. B* est
5–6 DE[2]. . . .autem *illeg. D*
6 *supra* et EZ *add. O* ex secunda propositione / *supra* totum ergo *add. O* ex IIIª huius / *post* autem *scr. OP* et
6–7 *supra* et EZ *add. O* ex prima
7–8 datam. . . .EZ[1] *om. P* / quantitatis. . . .Et *illeg. D*
8–9 compo[nenti. . . .converten]ti *illeg. D*
9 convertenti *add. mg. B*
10 quoniam: quia *B*
10–11 ad. . . .ad *illeg. D*
11 scilicet *om. B* / AG: AB *B*
12 ut[rumque. . . .data *illeg. D*
13 VII: 7ª *B*
13–14 utraque. . . .da]ta *illeg. D*
15–16 quantitatis. . . .GB[1] *illeg. D*
17 quanti[tatis. . . .proportio *illeg. D*
18 *post* AG *add. mg. O* ex premissa / *de* Data *add. mg. D* ex II elementi
19 AG: AB *B* / scilicet *om. B* / *post* GB *add. mg. O* per secundam / sectionum. . . .GB *illeg. D*
20 VIII: 8ª *B* / habentia: habencia *DP*
20–21 propo[rtionem[2]. . . . da]tam *illeg. D*

another, be conjoined. I say that the whole *AB* also has a given ratio to each of them, namely to *AG* and *GB*.

Let a given quantity *DE* be laid down. And since the ratio *AG*:*GB* is given, let a ratio equal to it, *DE*:*EZ*, be assigned. Then the ratio *DE*:*EZ* is given. Now *DE* is given. Therefore *EZ* is also given. Therefore the whole *DZ* is also given. Now each of *DE* and *EZ* are given. Therefore the ratio of the quantity *DZ* to each of *DE* and *EZ* is also given. And since *DE*: *EZ* = *AG*:*GB*, by composing, *DZ*:*EZ* = *AB*:*GB*. And by converting, *DZ*: *DE* = *AB*:*AG*. And since as *DZ* is to each of *DE* and *EZ*, so is *AB* to each of *AG* and GB, the ratio of the quantity *AB* to each of *AG* and *GB* is given.

VII. If a given quantity is divided in a given ratio, each of the segments will be given.

Let the given quantity *AB* be divided in the given ratio, namely, *AG*:*GB*. I say that each of *AG* and *GB* is given.

Since the ratio *AG*:*GB* is given, the ratio of the quantity *AB* to each of *AG* and *GB* is given. Now *AB* is given. Therefore each of the segments, namely, *AG* and *GB*, is also given.

VIII. Those which have a given ratio to the same, will also have a given ratio to each other.

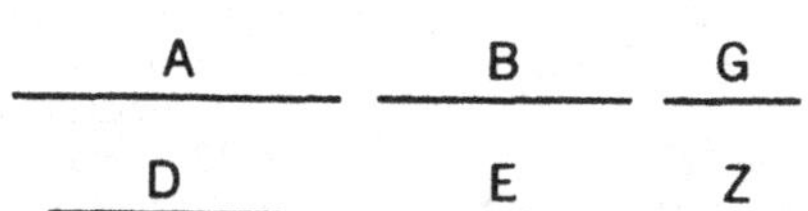

Habeat enim utrumque horum scilicet *A*, *G* ad *B* proportionem datam. Dico quod et *A* ad *G* proportionem habebit datam.

Esto enim data quantitas D. Et quoniam proportio quantitatis *A* ad *B* data est, ei eadem fiat quantitatis *D* ad *E*. At data *D*. Data ergo et *E*. Rursum quoniam proportio quantitatis *B* ad *G* data est, ei eadem fiat quantitatis *E* ad *Z*. At data *E*. Data ergo et *Z*. Est autem et *D* data. Proportio ergo quantitatis *D* ad *Z* est data. Et quoniam est sicut *A* ad *B* ita *D* ad *E* et ut *B* ad *G* ita *E* ad *Z*, equaliter ergo ut *A* ad *G* ita *D* ad *Z*. At proportio quantitatis *D* ad *Z* data. Proportio ergo et quantitatis *A* ad *G* data.

IX. Si due vel plures quantitates adinvicem proportionem habuerint datam habuerintque eedem quantitates ad aliquas alias quantitates proportiones datas, si non etiam easdem, et ille quantitates adinvicem proportiones habebunt datas.

Due enim vel plures quantitates ut *A*, *B*, *G* advinvicem proportionem habeant datam, habeantque eedem quantitates *A*, *B*, *G* ad alias aliquas quantitates verbi gratia *D*, *E*, *Z* proportiones datas, non easdem etiam. Dico quod *D*, *E*, *Z* quantitates adinvicem proportiones habebunt datas.

Quoniam enim proportio est quantitatis *A* ad *B* data quantitatisque *A* ad *D* proportio data, quantitatis ergo etiam *D* ad *B* proportio est data. Sed et quantitatis *B* ad *E* proportio est data. Quantitatis ergo etiam *D* ad *E* proportio est data. Rursum quoniam proportio quantitatis *B* ad *G* est

2 et *om. B* / ad. . . .habebit *illeg. D*
3 Esto: Est *P*
4 data. . . .fiat *illeg. D*
5 quantita[tis. . . .ea]dem *illeg. D*
6 *post* quantitatis *bis D* B ad G data est, ei eadem fiat quantitatis / *de* Z[2] *add. mg. D* ex II elementi / At. . . .et[1] *illeg. D*
7 *de* Proportio *add. mg. D* per primum
7–8 sicut. . . .ad[1] *illeg. D*
8 *de* equaliter *add. mg. D* per XXII V elementorum / *post* equaliter *add. mg. O* per XXII Vti geometrie
9 quantitatis[1]: quantitatum *B* / At . . . .Z *om. D* / et *om. B*
9–10 Proportio. . . .data *illeg. D*
11 IX *om. B* / habuerint *illeg. D*
12–13 ali[as. . . . propor]tiones *illeg. D*
13 adinvicem *tr. B post* proportiones *ln. 14*
13–14 adin[vicem proportiones *illeg. D*
15–16 adinvicem. . . .datam *illeg. D*
16 habeant: habent *B*
16–17 A. . . .Z *illeg. D*
18 quantitates. . . .datas *illeg. D* / proportiones *om. B*
19–20 data. . . .quantita]tis *illeg. D*
20 *post* etiam *add. mg. O* ex premissa / *de* est *add. mg. D* per 8
21 quantitatis: quantitas *P* / Quantitatis: Quantitas *P*
21–22 Sed. . . .proportio[2] *illeg. D*
22 quantitatis: quantitas *P* / est[2] *tr. B ante* quantitatis

Let each of A and G have a given ratio to B. I say that A will also have a given ratio to G.

Let a quantity D be given. And since the ratio $A:B$ is given, let a ratio equal to it, $D:E$, be made. Now D is given. Therefore E is also given. Further, since the ratio $B:G$ is given, let a ratio equal to it, $E:Z$, be made. Now E is given. Therefore Z is also given. Now D is also given. Therefore the ratio $D:Z$ is given. And since $A:B = D:E$ and $B:G = E:Z$, therefore, by equality, $A:G = D:Z$. Now the ratio $D:Z$ is given. Therefore the ratio $A:G$ is also given.

IX. If two or more quantities should have a given ratio to one another, and if the same quantities should have given ratios, though they are not the same, to some other quantities, these other quantities will also have given ratios to one another.

Let two or more quantities as A, B and G have a given ratio to one another, and let the same quantities A, B and G have given ratios, though they are not the same, to some other quantities, for example, D, E and Z. I say that the quantities D, E and Z will have given ratios to one another.

Since the ratio $A:B$ is given and the ratio $A:D$ is given, the ratio $D:B$ is also given. But the ratio $B:E$ is also given. Therefore the ratio $D:E$ is also given. Further, since the ratios $B:G$ and $B:E$ are given, the ratio $E:G$ is also given. But the ratio $G:Z$ is given. Therefore, the ratio $E:Z$ is also given. Hence, the quantities D, E and Z have a given ratio to one another.

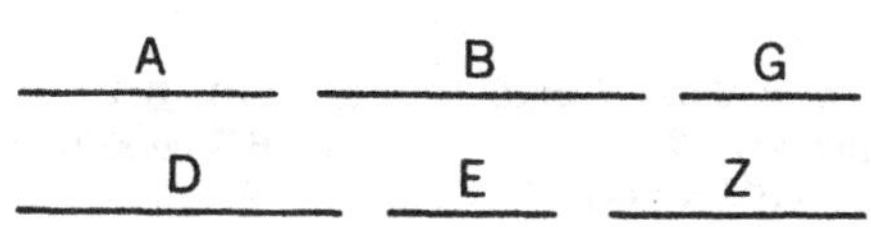

data quantitatisque *B* ad *E* proportio data, quantitatis ergo quoque *E* ad *G* proportio est data. Quantitatis vero *G* ad *Z* proportio est data. Quantitatis ergo etiam *E* ad *Z* proportio est data. Quantitates ergo *D*, *E*, *Z* adinvicem proportionem habent datam.

X. Si quantitas quantitate dato maior est quam in proportione, et coutrumque eadem dato maius est quam in proportione, et si coutrumque eadem dato maius est quam in proportione, et reliquum eadem vel dato maius est quam in proportione vel reliquum cum eo quod deinceps, ad quod alterum proportionem habet datam, datum est.

Quantitas enim *AB* quantitate *BG* dato maior esto quam in proportione. Dico quod coutrumque *AG* eadem *GB* dato maius est quam in proportione.

Quoniam enim *AB* quantitate *BG* dato maior est quam in proportione, auferatur data quantitas *AD*. Reliqui ergo *DB* ad *BG* proportio est data. Et componenti quantitatis *DG* ad *BG* proportio est data. Est autem data *AD*. Quantitas ergo *GA* quantitate *GB* dato maior est quam in proportione.

At rursum quantitas *AG* quantitate *GB* dato maior sit quam in proportione. Dico quod reliquum *AB* eadem *BG* vel dato maius est quam in proportione vel *AB* cum eo quod deinceps, ad quod *BG* proportionem
(59v) habet datam,/datum est.

1 proportio *tr. B post* data[2] / ergo *om. OP* / quoque *om. B*
1–2 quantitatisque. . . .data[1] *illeg. D*
2–3 ad. . . .data *illeg. D*
3 est *om. B*
3–4 ad[invicem. . . .datam *illeg. D*
5 X *om. B*
5–6 quan[titate. . . . coutrumque *illeg. D*
6 eadem: eodem *B*
6–7 ma[ius. . . . maius *illeg. D*
7–8 propo[rtione. . . . propor]tione *illeg. D*
8–9 reliquum. . . . datam *illeg. D*
10–14 est. . . . quantita]te *illeg. D*
11 dato *om. B*
11–17 maior. . . .AD *illeg. D*
12 GB: BG *O*
14 BG *corr. B ex* BD / dato *supra scr. B*
15 *supra* DB ad BG *add. O* per diffinitionem
16 *post* componenti *add. mg. O* per VI / quantitatis *om. B*
17 *supra* maior *add. O* ex diffinitione
17–19 ergo. . . .rur]sum *illeg. D* / est . . . .maior *om. P*
19 At: Et *B*
19–20 quantitas. . . .BG *illeg. D*
20 AB: AG *P* / eadem: eodem *OP* / BG: GB *P*
20–1(p.s.) maius. . . .est *illeg. D*
21 quod[2]: quam *B* / BG *corr. B ex* DG

X. If a quantity is greater than another quantity by a given quantity as in ratio [*i.e.* if the excess of a first quantity over the given quantity has a given ratio to the second quantity], the quantity composed of both [first and second quantities] is also greater than the same [second] quantity by a given quantity as in ratio [*i.e.* the excess of the sum of the first and second quantities over a given quantity has a given ratio to the second quantity], and if the quantity composed of both is greater than the same quantity by a given quantity as in ratio, then either the remaining [first quantity] is greater than the same [second] by a given quantity as in ratio [*i.e.* the excess of the first quantity over the given quantity has a given ratio to the second], or the remaining [first quantity] is given together with the following magnitude to which the second quantity has a given ratio.*

Let the quantity *AB* be greater than another quantity *BG* by a given quantity as in ratio [*i.e.* the excess of *AB* over a given quantity has a given ratio to *BG*]. I say that the composed quantity *AG* is greater than the same quantity *BG* by a given quantity as in ratio [*i.e.* the excess of *AG* over a given quantity also has a given ratio to *BG*].

Since *AB* is greater than *BG* by a given quantity as in ratio [*i.e.* since the excess of *AB* over a given quantity has a given ratio to *BG*], let the given quantity *AD* be taken away. Then the ratio of the remainder *DB* to *BG* is given. And, by composing, the ratio $DG:BG$ is also given. But the quantity *AD* is given. Therefore *GA* is greater than *GB* by a given quantity as in ratio [*i.e.* with *AD* given, $(GA - AD):BG$ is given].

Now further, let the quantity *AG* be greater than the quantity *GB* by a given quantity as in ratio [*i.e.* let the excess of *AG* over a given quantity have a given ratio to *GB*]. I say that either the remainder *AB* is greater than the same quantity *BG* by a given quantity as in ratio [*i.e.* either the excess of *AB* over a given quantity has a given ratio to *BG*] or *AB* is given with the quantity which follows to which *BG* has a given ratio.

* Since the clause "a quantity is greater than another quantity by a given quantity as in ratio," although rendering the Latin (and Greek) literally, is obscure, I have by using definition 11, added the precise meanings in brackets.

Quoniam enim *AG* quantitas quantitate *GB* dato maior est quam in proportione, auferatur data quantitas. Et data vel minor est quantitate *AB* vel maior.

Esto primum minor et sit *AD*. Reliqui ergo *DG* ad *GB* proportio est data. Dividenti ergo quantitatis *DB* ad *BG* proportio est data. Atqui data est *AD*. Quantitas ergo *AB* quantitate *BG* dato maior est quam in proportione.

Sed et datum maius esto quantitate *AB* et ponatur ei equalis *AE*. Proportio ergo reliqui *EG* ad *GB* est data. Quare et contra quantitatis *BG* ad *EG* proportio est data, et convertenti quantitatis *BG* ad *BE* proportio est data. Atqui *EB* cum *BA* datum, totum enim *AE* datum est. Quantitas ergo *BA* cum eo quod deinceps, ad quod *BG* proportionem habet datam, data est.

XI. Si quantitas quantitate dato maior est quam in proportione, eadem etiam coutroque dato maior erit quam in proportione. Et si eadem coutroque dato maior est quam in proportione, eadem etiam reliquo dato maior erit quam in proportione.

Quantitas enim *AB* quantitate *BG* dato maior esto quam in proportione. Dico quod et quantitate *AG* dato maior est quam in proportione.

Quoniam enim quantitas *AB* quantitate *BG* dato maior est quam in pro-

1 quantitate: quantitatis *P*
2 quantitate *om. B*
2–4 quantitas. . . .primum *illeg. D*
4 Reliqui: Relinqui *B* / *de* Reliqui *add. mg. D* per diffinitionem / *supra* DG ad GB *add. O* ex diffinitione / *post* est *scr. B* etiam
4–5 ergo. . . .DB *illeg. D*
5 *post* DB *add. mg. O* per V et VIII huius / Atqui: Sed *B*
6 est[1]: et *B*
5–6 At[qui. . . .maior *illeg. D*
6 *post* dato *add. mg. O* per diffinitionem
8 datum. . . .equalis *illeg. D*
9 reliqui: reliquum *B* / *supra* EG ad GB *add. O* per diffinitionem / *post* ad *add. mg. D* per eadem / *post* BG *add. mg. O* per diffinitionem date proportionis et per V librum / GB . . . .ad *illeg. D*
10 est[1] *om. P* / et: etiam *P* / *supra* convertenti *add. O* per V huius
10–11 BG. . . .cum *illeg. D*
11 enim *correxi ex* ergo *BDOP* / est *om. B*
12 BA. . . .BG *illeg. D* / quod[2]: quem *B*
14 XI *om. B* / Si. . . . quantitate *illeg. D*
15 dato. . . . quam *illeg. D*
16–17 propor[tione. . . . reli]quo *illeg. D*
18 maior. . . .proportione *illeg. D*
19 quantitate: quantitas *B*
20 AB. . . .dato *illeg. D*

Since the quantity *AG* is greater than the quantity *GB* by a given quantity as in ratio [*i.e.* since the excess of *AG* over a given quantity has a given ratio to *GB*], let the given quantity be taken away [from *AG*]. And the given quantity is less than the quantity *AB* or greater than it.

Let it, first, be less [than *AB*] and be *AD*. Then the ratio of the remainder *DG* to *BG* is given. Therefore, by dividing, the ratio *DB*:*BG* is given. But the quantity *AD* is given. Therefore the quantity *AB* is greater than the quantity *BG* by a given quantity as in ratio [*i.e.* the excess of *AB* over a given quantity has a given ratio to *BG*].

But, let the given quantity be greater than the quantity *AB* and let the quantity equal to it, *AE*, be posed. Then the ratio of the remainder *EG* to *GB* is given. Therefore, conversely, the ratio *BG*:*EG* is given and, by converting, the ratio *BG*:*BE* is also given. But the sum of *EB* and *BA* is given, for the whole *AE* is given. Therefore *BA* is given together with that which follows [*i.e. BE*], to which *BG* has a given ratio.

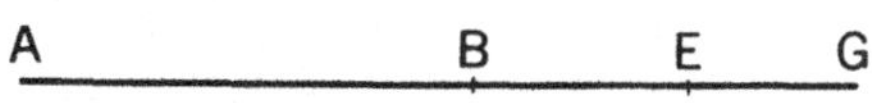

XI. If a quantity is greater than a quantity by a given quantity as in ratio [*i.e.* if the excess of a quantity over a given quantity has a given ratio to another quantity], the same [first] quantity will also be greater than the quantity compound of them by a given quantity as in ratio [*i.e.* the excess of the same first quantity over a given quantity will have a given ratio to the sum of both quantities]. And if the same [first] quantity is greater than the quantity compound of them by a given quantity as in ratio [*i.e.* if the excess of the first quantity over a given quantity has a given ratio to the sum of both quantities], the same will also be greater than the remainder by a given quantity as in ratio [*i.e.* the excess of the first over a given quantity has a given ratio to the second].

portione, auferatur data quantitas *AD*. Reliqui ergo *DB* ad *BG* proportio est data. Econtra ergo et componenti proportio ergo *GD* ad *DB* data. Et eadem ei fiat quantitatis *AD* ad *DE*. Proportio ergo etiam quantitatis *AD* ad *DE* data. Atqui data *AD*. Data ergo et *DE*. Quare et reliquum *EA* datum est. Et est etiam totius *AG* ad totum *EB* proportio data. Quare et *EB* ad *AG* proportio est data. Atqui data est *AE*. Quantitas ergo *BA* quantitate *AG* dato maior est quam in proportione.

Sed et *BA* coutroque *AG* dato maior esto quam in proportione. Dico quod eadem *BA* etiam reliquo *BG* dato maior erit quam in proportione.

Quoniam enim quantitas *AB* quantitate *AG* dato maior est quam in proportione, auferatur data quantitas *AE*. Reliqui ergo *EB* ad *AG* proportio est data. Quare et *AG* ad *EB* proportio est data. Et eadem ei fiat quantitatis *AD* ad *ED*. Itaque et *DA* ad *ED* proportio est data, et convertenti *DA* ad *AE* proportio data, et econtra *EA* ad *AD* proportio est data. Atqui data *AE*. Data ergo et *ED*. Totum ergo etiam *AD* datum est. Et quoniam totius *AG* ad totum *EB* proportio est data, proportio ergo *AD* ad *DE* est data. Et reliqui ergo *GD* ad reliquum *DB* proportio est data. Et dividenti *GB* ad *BD* proportio est data. Quare et *DB* ad *BG* proportio est data. Et quoniam data est *DA*, *AB* ergo *BG* dato maior est quam in proportione.

1 *de* Reliqui *add. mg. D* per diffinitione / ergo: igitur *B* / BG: GB *B*
1–2 DB. . . .est *illeg. D*
2 est data: data est *B* / *de* Econtra *add. mg. D* per diffinitionem date / *post* Econtra *add. mg. O* per diffinitionem date proportionis et V librum / *supra* GD ad DB *add. mg. O* per VI huius
3 eadem. . . .quantitatis[1] *illeg. D* / eadem *tr. B post* ei / *post* ei *scr. et del. P* / AD ad DE: DE ad AD *B* / etiam *om. B*
4 Data. . . .reli]quum *illeg. D* / *supra* et DE *add. O* ex secunda / *post* Quare et *add. mg. O* per IIII
5 *supra* AG ad totum *add. O* per XIII, V elementorum / *de* totum *add. mg. D* per 12, V euclidis
5–6 et. . . .est[1] *illeg. D*
6 *post* EB ad AG *add. mg. O* ex diffinitione date proportionis et V libro elementorum / de Quantitas *add. mg. D* ex diffinitione / BA: DA *B*
7 quantitate: quantitas *B* / *supra* maior *add. O* ex diffinitione
7–8 est. . . .et *illeg. D*
8 et *om. B* / AG *om. B*
9 etiam. . . .maior *illeg. D*
10–11 est. . . .aufer]atur *illeg. D*
12 et. . . .est *illeg. D* / et *om. B*
13 ED[2]: DE *B*
14 DA: DEA *B* / *post* DA *add. mg. O* per V huius / *ante* data *scr. B* est
14–16 et. . . .data *om. B*
15 Data. . . .Totum *illeg. D*
16 totum *corr. DO ex* datum
16–17 ergo. . . .Et[1] *illeg. D*
17 *de* reliqui *add. mg. O* per XIX V elementorum *et D* per 19, 5 euclidis / GD: DG *B*
18 GB: GD *B* / BD. . . .et *illeg. D* / est[2] *tr. B post* data[1] *ln. 19*
19–1(p.s.) quam. . . . pri]ma *illeg. D*

Let the quantity *AB* be greater than the quantity *BG* by a given quantity as in ratio [*i.e.* let the excess of *AB* over a given quantity have a given ratio to *BG*]. I say that it [the quantity *AB*] is also greater than the quantity *AG* by a given quantity as in ratio [*i.e.* the excess of *AB* over a given quantity has a given ratio to *AG*].

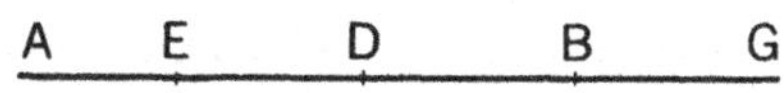

Since the quantity *AB* is greater than the quantity *BG* by a given quantity as in ratio [*i.e.* the excess of *AB* over a given quantity has a given ratio to *BG*], let the given quantity *AD* be taken [from *AB*]. Then the ratio of the remainder *DB* to *BG* is given. Therefore, by inversion and by composition, the ratio *GD*:*DB* is also given. And let the same be done with the ratio *AD*:*DE*. Then the ratio *AD*:*DE* is also given. Now *AD* is given. Therefore *DE* is also given. Consequently the remainder *EA* is also given. And the ratio of the whole *AG* to the whole *EB* is also given. Consequently the ratio *EB*:*AG* is given. Now *AE* is given. Therefore the quantity *BA* is greater than *A*G by a given quantity [*EA*] as in ratio [*i.e.* with *EA* given, the ratio of quantities $(BA-EA):AG$ is given].

But let *BA* be greater than the composed quantity *AG* by a given quantity as in ratio. I say that the same *BA* will also be greater than the remainder *BG* by a given quantity as in ratio [*i.e.* the excess of *AB* over a given quantity will also have a given ratio to *BG*].

Since the quantity *AB* is the quantity *AG* by a given quantity as in ratio [*i.e.* since the excess of *AB* over a given quantity has a given ratio to *AG*], let the given quantity *AE* be taken [from *AB*]. Then the ratio of the remainder *EB* to *AG* is given. Consequently the ratio *AG*:*EB* is also given. And let the same thing be done with the ratio of *AD* to *ED*. Then the ratio *DA*:*ED* is also given and, by converting, the ratio *DA*:*AE* is also given and, by inversion, the ratio *EA*:*AD* is also given. Now *AE* is given. Therefore *ED* is also given. Hence, the whole *AD* is also given. Since the ratio of the whole *AG* to the whole *EB* is given, the ratio *AD*:*DE* is given. And therefore, the ratio of the remainder *GD* to the remainder *DB* is given. And by dividing, the ratio *GB*:*BD* is given. Consequently, the ratio *DB*:*BG* is given. Since *DA* is given, *AB* is greater than *BG* by a given quantity [*AD*] as in ratio [*i.e.* the excess of *AB* over the given quantity *AD* has a given ratio to *BG*].

XII. Si fuerint tres quantitates et prima quidem cum secunda est data, et secunda cum tertia data prima tertie vel equalis erit vel altera altera dato maior.

Sint tres quantitates *AB*, *BG*, *GD*. Sitque *AB* cum *BG* data *AG* et *BG* cum *GD* data esto *BD*. Dico quod *AB* quantitas quantitati *GD* vel equalis est vel altera altera dato maior est.

Qunoniam enim data est utraque *AG*, *BD*, data vel equalia sunt vel inequalia. Sint primum equalia. Equalis ergo *AG* quantitas quantitati *BD*. Commune auferatur *BG*. Reliquum ergo *AB* reliquo *GD* equale est.

Non autem sint equalia. Sed esto maior *AG* quantitas quantitate *BD*, et ponatur equalis quantitas *BD* quantitati *GE*. At data *BD*. Data ergo et *GE*. Estque etiam totum *AG* datum. Et reliquum ergo *AE* datum est. Et quoniam equalis est *EG* quantitas quantitati *BD*, commune auferatur *BG*. Reliquum ergo *BE* reliquo *GD* equale est. Et est datum *AE*. Quantitas ergo *AB* quantitate *GD* dato maior est.

XIII. Si fuerint tres quantitates et prima ad secundam proportionem habuerit datam, secundaque tertia maior erit dato quam in proportione, prima etiam tertia dato maior erit quam in proportione.

Sint tres quantitates *AB*, *GD*, *E* et *AB* ad *GD* proportionem habeat datam. Quantitasque *GD* quantitate *E* dato maior sit quam in proportione.
(60r) Dico quod et *AB* quantitas quantitate *E* dato maior/est quam in proportione.

1 XII: 12ª *B*
2 cum. . . . erit *illeg. D* / tertia: tercia *BOP hic et ubique* / tertie: tercie *BOP*
3 *post* maior *scr. B* est
4 Sint. . . .et *illeg. D* / Sitque: Sit *B* / AG *del. B*
5 GD[1]: DG *B* / BD *del. B*
5–6 quod. . . .mai]or *illeg. D*
7 sunt *om. B*
7–8 AG. . . .A]G *illeg. D*
8 *post* quantitas *supra scr. B* est
9–10 auferatur. . . .si]nt *illeg. D*
10 sint: sunt *B*
10–11 quantitas. . . .At *illeg. D*
12 est *om. B* / Et[2] *supra scr. B*
12–13 Est[que. . . .EG *illeg. D*
13–14 au[feratur. . . .Quanti]tas *illeg. D*
16 XIII: 13ª *B* / tres: 3 *B* / Si. . . . proportionem *illeg. D*
17–18 datam. . . . prima *illeg. D*
18 prima: 1ª *B* / etiam: et *scr. et del. B* / tertia: 3ª *B*
18–19 maior. . . . ad *illeg. D*
19 habeat: habuit *B*
19–21 habeat. . . .quantitate *illeg. D*
21–2(p.s.) est. . . .Reliqui *illeg. D*

XII. If there are three quantities and the first with the second is given and also the second with the third is given, either the first will be equal to the third or one will be greater than the other by a given quantity.

Let there be three quantities *AB*, *BG* and *GD* and let *AB* with *BG* [= *AB* + *BG*] be given as *AG*, and *BG* and *GD* [= *BG* + *GD*] as *BD*. I say that either the quantity *AB* is equal to the quantity *GD* or one is greater than the other by a given quantity.

Since each of *AG* and *BD* are given, they are either equal to each other or unequal. Let them, first, be equal. Then the quantity *AG* is equal to the quantity *BD*. Let the common quantity *BG* be taken away. Then the remainder *AB* is equal to the remainder *GD*.

Now, let them be not equal, but let the quantity *AG* be greater than the quantity *BD* and let the quantity *GE* be put equal to the quantity *BD*. Now *BD* is given. Therefore *GE* is also given. And the whole *AG* is also given. Therefore the remainder *AE* is also given. Since the quantity *EG* is equal to the quantity *BD*, let the common quantity *BG* be taken away. Then the remainder *BE* is equal to the remainder *GD*. *AE* is also given. Therefore the quantity *AB* is greater than the quantity *GD* by a given quantity [*AE*].

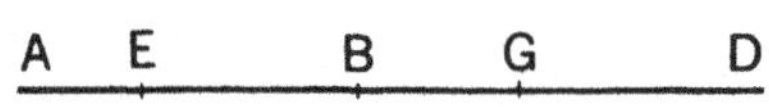

XIII. If there are three quantities and the first has a given ratio to the second and the second is greater than the third by a given quantity as in ratio [*i.e.* the excess of the second over a given quantity has a given ratio to the third], the first will also be greater than the third by a given quantity as in ratio [*i.e.* the excess of the first over a given quantity has a given ratio to the third].

Let the three quantities be *AB*, *GD* and *E* and let *AB* have a given ratio to *GD*. Let the quantity *GD* be greater than the quantity *E* by a quantity as in ratio [*i.e.* let the excess of *GD* over a given quantity have a given ratio to *E*]. I say that the quantity *AB* is also greater than the quantity *E* by a given quantity as in ratio [*i.e.* the excess of *AB* over a given quantity has a given ratio to *E*].

Quoniam enim *GD* quantitas quantitate *E* dato maior est quam in proportione, auferatur data quantitas *GZ*. Reliqui ergo *DZ* ad *E* proportio est data. Et quoniam proportio est data quantitatis *AB* ad *GD*, eadem ei fiat quantitatis *AI* ad *GZ*. Proportio ergo etiam quantitatis *AI* ad *GZ* data. Atqui data etiam *GZ*. Data ergo et *AI*. Et reliqui *IB* ad reliquum *DZ* proportio est data. Et *DZ* ad *E* proportio est data. Et *IB* ergo ad *E* proportio est data. Atqui data est *AI*. Quantitas ergo *AB* quantitate *E* dato maior est quam in proportione.

XIIII. Si due quantitates adinvicem proportionem habent datam apponaturque utrique earum data quantitas, tota adinvicem proportionem habebunt datam vel alterum altero dato maius est quam in proportione.

Due enim quantitates *AB*, *GD* adinvicem proportionem habeant datam. Et apponatur utrique earum data quantitas *AE* et *GZ*. Dico quod tota *EB* et *ZD* adinvicem vel proportionem habebunt datam vel alterum altero dato maius est quam in proportione.

Quoniam enim datum est utrumque *EA* et *ZG*, proportio ergo quantitatis *EA* ad *ZG* data. Et si eadem est quantitatis *AB* ad *GD* erit, etiam totius *EB* ad totum *ZD* proportio data. Non sit autem eadem et fiat sicut *AB* ad

1 *post* quantitate *bis D* E dato maior est quam in proportione.] Quoniam enim GD [quantitas quantitate
2 quantitas *om. B* / GZ *corr. B ex* DZ
2–4 ad. . . . quanti]tatis[1] *illeg. D*
4 GZ: ZG *B*
4–5 AI[1]. . . .reliqui *illeg. D*
5 Atqui: Atque *B* / GZ: ZG *B* / IB: BB *B* / *de* IB *add. mg. D* per 19, 5 euclidis / reliquum: relicuum *B* / *de* reliquum *add. mg. O* per XIX V[ti] elementorum
5–7 reliquum. . data[1] *illeg. D*
6 est[2] om. *OP*
6–7 Et[2]. . . .data[1] *om. B*
7 *supra* est data *add.* O ex VIII / Atqui: Sed *B* / est[2] *om. B*
9 XIIII: 14[a] *B* / *comment. illeg. scr. mg. manu rec. et del. O*
9–10 adinvicem. . . . apponaturque *illeg. D*
10–11 quan[titas. . . . dat]am *illeg. D*
11 maius *corr. O ex* maior
11–13 mai[us. . . . adinvicem *illeg. D*
14 GZ: ZG *B*
14–15 earum. . . .ZD *illeg. D*
15 habebunt: habent *BP*
15–17 altero. . . .enim *illeg. D*
17 EA: AE *B* / et *om. BP* / ZG: GZ *BO*
18 *supra* EA ad ZG *add. O* per primam huius / ZG. . . .quantitat]is *illeg. D*
19 *supra* ad totum ZD *add. O* per XIII V[ti] elementorum / sit. . . . AB *illeg. D*

Since the quantity *GD* is greater than the quantity *E* by a given quantity as in ratio [*i.e.* since the excess of *GD* over a given quantity has a given ratio to *E*], let the given quantity *GZ* be taken away. Then the ratio of the remainder *DZ* to *E* is given. Since the ratio *AB*:*GD* is given, let the same ratio as it be made of *AI* to *GZ*. Then the ratio *AI*:*GZ* is also given. Now *GZ* is also given. Therefore *AI* is also given. And the ratio of the remainder *IB* to the remainder *ZD* is also given. The ratio *DZ*:*E* is given. Therefore the ratio *IB*:*E* is given. But *AI* is given. Therefore the quantity *AB* is greater than the quantity *E* by a given quantity [*AI*] as in ratio [*i.e.* the excess of *AB* over the given quantity *AI* has a given ratio to *E*].

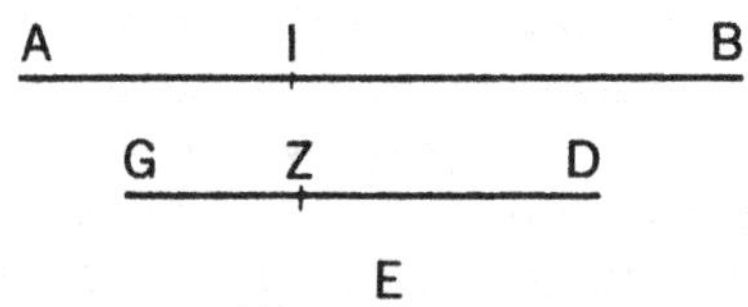

XIV. If two quantities have a given ratio to each other and to each of them there is added a given quantity, either the wholes will have a given ratio to each other or the one is greater than the other by a given quantity as in ratio.

Let two quantities *AB* and *GD* have a given ratio to each other. And let given quantities *AE* and *GZ* be added [respectively] to each of them [*AB*, *GD*]. I say that either the wholes *EB* and *ZD* will have a given ratio to each other or the one is greater than the other by a given quantity as in ratio.

Since each of the quantities *EA* and *ZG* is given, the ratio *EA*:*ZG* is given. And if that ratio is the same as the ratio *AB*:*GD*, the ratio of the whole *EB* to the whole *ZD* will be given.

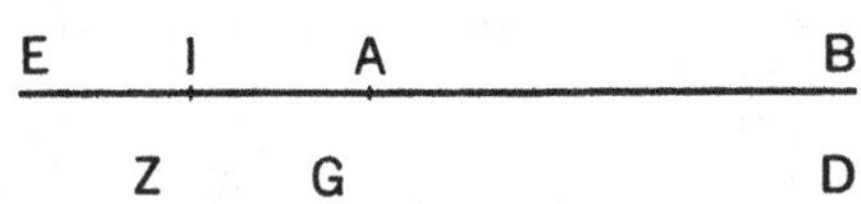

GD ita *IA* ad *ZG*. Proportio ergo etiam quantitatis *IA* ad *ZG* data. Atqui data *ZG*. Data ergo etiam *IA*. Est quoque et *EA* data. Et reliquum ergo *EI* datum est. Et quoniam sicut *AB* ad *GD* ita *IA* ad *ZG*, proportio ergo etiam quantitatis *IB* ad *ZD* data. Atqui data est *EI*. Quantitas ergo *EB* quantitate *ZD* dato maior est quam in proportione.

XV. Si due quantitates adinvicem proportionem habuerint datam, auferaturque ab utraque earum data quantitas, reliqua adinvicem proportionem habebunt datam vel alterum altero dato maius est quam in proportione.

Due enim quantitates *AB* et *GD* adinvicem proportionem habeant datam, auferaturque ab utraque earum data quantitas ab *AB EA* et ab *GD GZ*. Dico quod reliqua *EB* et *ZD* adinvicem vel proportionem habebunt datam vel alterum altero dato maius est quam in proportione.

Quoniam enim utrumque *AE*, *GZ* datum est, proportio ergo *AE* ad *GZ* data. Et si eadem est quantitatis *AB* ad *GD*, erit et reliqui *EB* ad reliquum *ZD* proportio data.

Non sit autem eadem, fiatque sicut *AB* ad *GD* ita *AI* ad *GZ*. Atqui proportio quantitatis *AB* ad *GD* data. Proportio ergo et quantitatis *AI* ad *GZ* data, datumque *GZ*. Datum ergo et *AI*. Estque et *AE* datum. Et reliquum ergo *EI* datum est. Et quoniam sicut *AB* ad *GD* ita *AI* ad *GZ*, reliqui ergo *IB* ad reliquum *ZD* proportio data. Et est datum *EI*. Quantitas ergo *EB* quantitate *ZD* dato maior est quam in proportione.

1 etiam *om. B*
2 data[1]....Est *illeg. D* / ZG:GZ *B* / etiam: et *B* / EA:AE *B*
3 EI:IE *B* / *post* quoniam *scr. DOP* est / ita....ergo *illeg. D* / ZG:GZ *OP*
4 etiam *om. B* / *post* IB *scr. et del. B* quantitate ZD dato maior est / Atqui: Sed *B* / est: et *D*
5 ZD....proportione *illeg. D*
6 XV: 15[a] *B*
6–7 datam, au]feraturque *illeg. D*
7 reliqua: tota *B*
7–8 ad[invicem. propor]tionem *illeg. D*
9 quam in proportione *illeg. D*
10 GD:DG *B*
11 auferaturque....earum *illeg. D* / EA:AE *B*
12 adin[vicem....habebunt *illeg. D*
13 dato maius est: est maius dato *B*
14 utrumque: utraque *B* / GZ....ergo *illeg. D* / *post* AE[2] *bis D* GZ datam est, proportio ergo AE
15 AB....reliqui *illeg. D* / reliqui: reliquum *B*
17 *de* sit autem *add. mg. D* quia eadem queretur ex 5, 19 / fiatque: sed fiat *B* / Atqui: Sed *B*
17–18 ad[1]....d]ata....*illeg. D*
18 quantitatis[1]: quantitatem *B* / AB ad GD: GD ad AB *B*
19 data....GZ *illeg. D*
20 AB....ad *illeg. D*
22 ergo....proportione *illeg. D*

Let that ratio be not the same [as $AB:GB$] and let it be that $AB:GD = IA:ZG$. Then the ratio $IA:ZG$ is given. Now ZG is given. Therefore IA is also given. And EA is also given. Therefore the remainder EI is also given. Since $AB:GD = IA:ZG$, the ratio $IB:ZD$ is also given. Now EI is given. Therefore the quantity EB is greater than the quantity ZD by a given quantity [EI] as in ratio [*i.e.* the excess of EB over the given quantity EI has a given ratio to ZD].

XV. If two quantities have a given ratio to each other and a given quantity is taken away from each of them, the remainders will have a given ratio to each other or the one is greater than the other by a given quantity as in ratio.

Let the quantities AB and GD have a given ratio to each other and let a given quantity be taken away from each of them, namely EA from AB, GZ from GD. I say that either the remainders EB and ZD will have a given ratio to each other or the one is greater than the other by a given quantity as in ratio [*i.e.* the excess of the one over a given quantity has a given ratio to the other].

Since each of AE and GZ is given, the ratio $AE:GZ$ is given. If that ratio will be the same as the ratio $AB:GD$, the ratio of the remainder EB to the remainder ZD is also given.

But let the ratio be not the same and let it be $AB:GD = AI:GZ$. Now $AB:GD$ is given. Therefore $AI:GZ$ is also given. And GZ is given. Therefore AI is given, and AE is also given. Therefore the remainder EI is also given. Since $AB:GD = AI:GZ$, the ratio of the remainder IB to the remainder ZD is given. And EI is given. Therefore the quantity EB is greater than ZD by a given quantity [EI] as in ratio [*i.e.* the excess of EB over the given quantity EI has a given ratio to ZD].

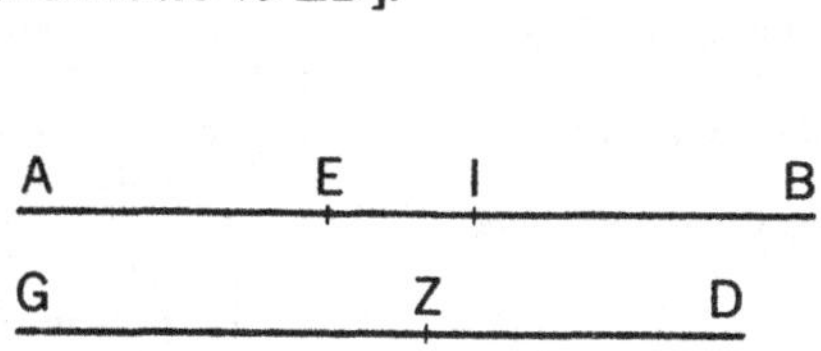

XVI. Si due quantitates adinvicem proportionem habeant datam, et ab una quidem earum data quantitas auferatur, alteri vero earumdem data quantitas apponatur, totum reliquo dato maius est quam in proportione.

Due enim quantitates *AB*, *GD* proportionem habeant datam et a quantitate *GD* data quantitas *GE* auferatur. Quantitati vero *AB* data quantitas *ZA* apponatur. Dico quod totum *ZB* reliquo *ED* dato maius est quam in proportione.

Quoniam enim proportio est quantitatis *AB* ad *GD* data, eadem ei fiat quantitatis *AI* ad *GE*. Proportio ergo etiam quantitatis *AI* ad *GE* data. Atqui data *GE*. Data ergo est *AI*. Estque data et *AZ*. Totum ergo *ZI* datum est. Et quoniam sicut *AB* ad *GD* ita *AI* ad *GE*, et reliqui *IB* ad reliquum *ED* proportio est data. Et est data *IZ*. Quantitas ergo *ZB* quantitate *ED* dato maior est quam in proportione.

XVII. Si fuerint tres quantitates et prima secunda dato maior fuerit
(60v) quam in proportione, tertia quoque eadem dato maior fuerit/quam in proportione, prima ad tertiam vel proportionem habebit datam vel altera altera dato maior erit quam in proportione.

Sint tres quantitates *AB*, *G*, *DE* et utraque *AB*, *DE* quantitate *G* dato maior sit quam in proportione. Dico quod *AB* et *DE* vel adinvicem proportionem habent datam vel altera altera dato maior est quam in proportione.

1 XVI *om. B*
1–2 datam. . . . u]na *illeg. D*
2–3 aufe[ratur . . . quanti]tas *illeg. D*
3–4 reliqu[o . . . proportione *illeg. D*
5–6 habeant. . . .Quantita]ti *illeg. D*
7 dato *om. B*
7–8 apponatur. . . .proportione *illeg. D*
9 est *om. P*
9–10 AB. . . .quanti]tatis[2] *illeg. D*
10 AI ad GE[1]: GE ad AI *B* / etiam *om. P*
11 Atqui. . . .AI: Ergo et da (?) *B*
11–12 data[1]. . . .quoniam *illeg. D*
12–13 et. . . .Quanti]tas *illeg. D*
13 est data[2]: data est *B*
14 da[to. . . .proportione *illeg. D*
15 XVII: 17[a] *B* / tres: 3 *B* / prima: 1[a] *B* / *post* prima *scr. BP* et
15–16 quantitates. . . . ter]tia *illeg. D*
16 tertia: 3[a] *B*
16–17 dato. . . . pro]portionem *illeg. D*
17 prima: 1[a] *B* / tertiam: 3[a] *B* / proportionem *tr. B post* habebit
17–18 datam. . . . proportione *illeg. D*
18 dato *om. B*
19 tres: 3 *B* / AB, DE: DE, AB *B* / dato *om. B*
19–20 quanti[tates. . . .quod *illeg. D*
20–1(p.s.) adinvicem. . . .enim *illeg. D*

XVI. If two quantities have a given ratio to each other and from one of them there be taken away a given quantity and to the other there be added a given quantity, then the whole is greater than the remainder by a given quantity as in ratio.

Let two quantities *AB* and *GD* have a given ratio to each other, and from *GD* let the given quantity *GE* be taken, and to *AB* let the given quantity *ZA* be added. I say that the whole *ZB* is greater than the remainder *ED* by a given quantity as in ratio [*i.e.* the excess of *ZB* over a given quantity has a given ratio to *ED*].

Since the ratio *AB*:*GD* is given, let the same ratio as it be made of *AI* to *GE*. Therefore the ratio *AI*:*GE* is also given. Now *GE* is given. Therefore *AI* is given. And *AZ* is also given. Therefore the whole *ZI* is given. Since *AB*:*GD* = *AI*:*GE*, the ratio of the remainder *IB* to the remainder *ED* is also given. *IZ* is also given. Therefore the quantity *ZB* is greater than

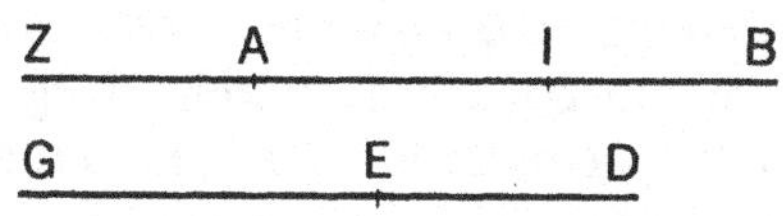

quantity *ED* by a given quantity [*IZ*] as in ratio [*i.e.* the excess of the quantity *ZB* over the given quantity *IZ* has a given ratio to the quantity *ED*].

XVII. If there are three quantities and if the first is greater than the second by a given quantity as in ratio and if the third is also greater than the same [second] as in ratio, then the first will have to the third either a given ratio, or else the one will be greater than the other by a given quantity as in ratio.

Let the three quantities be *AB*, *G* and *DE* and let each of *AB* and *DE* be greater than *G* by a given quantity as in ratio [*i.e.* let the excess of each of *AB* and *DE* over a given quantity have a given ratio to *G*]. I say that *AB* and *DE* either have a given ratio to each other or else the one is greater than the other by a given quantity as in ratio [*i.e.* the excess of one over a given quantity has a given ratio to the other].

Quoniam enim *DE* quantitas quantitate *G* dato maior est quam in proportione, auferatur data quantitas *DI*. Reliqui ergo *IE* ad *G* proportio est data. Per eadem vero et *ZB* ad *G* proportio est data. Et *ZB* ergo ad *IE* proportio est data. Et apponantur eisdem date quantitates *AZ*, *DI*. Tota igitur *AB* et *DE* adinvicem vel proportionem habent datam vel alterum altero dato maius est quam in proportione.

XVIII. Si fuerint tres quantitates, unaque earum utraque reliquarum dato maior fuerit quam in proportione, relique due adinvicem vel proportionem habebunt datam vel altera altera dato maior est quam in proportione.

Sint tres quantitates *AB*, *GD*, *EZ*, unaque earum *GD* utraque reliquarum *AB* et *EZ* quantitatum dato maior sit quam in proportione. Dico quod *AB* ad *EZ* vel proportionem habet datam vel altera altera dato maior est quam in proportione.

Quoniam enim quantitas *GD* quantitate *AB* dato maior est quam in proportione, auferatur data quantitas *GI*. Reliqui ergo *ID* ad *AB* proportio est data. Eadem ei fiat quantitatis *GI* ad *AT*. Proportio ergo et quantitatis *GI* ad *AT* data. Data autem *GI*. Data ergo et *AT*. Et totius *GD* ad totum *TB* proportio est data. Rursum quoniam quantitas *GD* quantitate *EZ* dato maior est quam in proportione, auferatur data quantitas *GK*. Reliqui ergo *KD* ad *EZ* proportio est data. Eadem ei fiat quantitatis *GK* ad *LE*. Proportio ergo et quantitatis *GK* ad *LE* data. Data autem *GK*. Data ergo et *LE*. Et totius *GD* ad totum *LZ* proportio est data. Proportio

1 enim *om. B* / dato *om. B*
1–2 quantitate. . . .IE *illeg. D*
3–4 propor[tio. . . .ap]ponantur *illeg. D*
4 *supra* data *add. O* ex VIII
4–6 eis[dem. . . .proportione *illeg. D*
5 AB:AG *B*
6 *post* dato *add. mg. O* ex XIIII^a
7 XVIII: 18^a *B* / tres: 3 *B*
7–8 fue[rint. . . . relique *illeg. D*
9 habebunt. . . . quam *illeg. D*
11 tres: 3 *B*
11–12 Sint. . . .reliquarum *illeg. D*
12–13 maior. . . .proportionem *illeg. D*
13–15 da[tam. . . .quantitas *illeg. D*
15 dato: da *B*
15–16 Quoniam. . . .proportione *om. P*
16 proportione. . . .GI *illeg. D* / Reliqui: relique *B* / ID:DI *B*
17 ei. . . .ergo *illeg. D* / quantitatis: quantitas *B* / et: etiam *B*
18 *de* $Data^2$ *add. mg. D* ex 2^a
18–19 AT^2. . . .proportio *illeg. D*
19 quoniam *om. B*
20 da[to. . . .proportione *illeg. D*
21–22 Reliqui. . . .GK^2 *om. B*
22 Proportio. . . .data *add. mg. D*
22–23 autem. . . .Et *illeg. D*

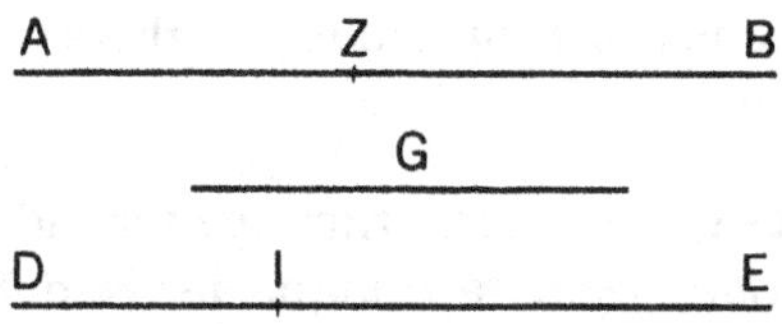

Since *DE* is greater than *G* by a given quantity *DI* as in ratio [*i.e.* the excess of *DE* over a given quantity has a given ratio to *G*], let a given quantity *DI* be taken away [from *DE*]. Then the ratio of the remainder *IE* to *G* is given. By the same argument, *ZB*:*G* is also given. Therefore *ZB*:*IE* is also given. And to them [*ZB*,*IE*] let the given quantities *AZ* and *DI* be added [respectively]. Then the wholes *AB* and *DE* have either a given ratio, or else the one is greater than the other by a given quantity as in ratio [*i.e.* the excess of the one over a given quantity has a given ratio to the other].

XVIII. If there are three quantities and one of them is greater than each of the remainders by a given quantity as in ratio, the two remainders will either have a given ratio to each other or else the one is greater than the other by a given quantity as in ratio.

Let the three quantities be *AB*, *GD* and *EZ* and let one of them, *GD*, be greater than either of the remaining quantities *AB* and *EZ* by a given quantity as in ratio [*i.e.* let the excess of *GD* over a given quantity have a given ratio to either *AB* or *EZ*]. I say that *AB* has either a given ratio to *EZ* or else the one is greater than the other by a given quantity as in ratio [*i.e.* the excess of the one over a given quantity has a given ratio to the other].

Since *GD* is greater than *AB* by a given quantity as in ratio [*i.e.* since the excess of *GD* over a given quantity has a given ratio to *AB*], let the given quantity *GI* be taken away. Then the ratio of the remainder *ID* to *AB* is given. Let the same ratio be made of *GI* to *AT*. Therefore *GI*:*AT* is also given. But *GI* is given. Therefore *AT* is also given. The ratio of the whole *GD* to the whole *TB* is also given. Further, since *GD* is greater than *EZ* by a given quantity as in ratio [*i.e.* since the excess of *GD* over a given quantity has a given ratio to *EZ*], let the given quantity *GK* be taken away. Then the ratio of the remainder *KD* to *EZ* is given. In the same way, let

quoque *GD* ad *TB* est data. Et *TB* ergo ad *LZ* proportio data. Et auferantur *AB* eis date quantitates *TA*, *LE*. Quantitates ergo *AB* et *EZ* vel adinvicem proportionem habent datam vel altera altera dato maior est quam in proportione.

XIX. Si fuerint tres quantitates, primaque secunda dato maior fuerit quam in proportione, secunda quoque tertia dato maior quam in proportione, prima etiam tertia dato maior erit quam in proportione.

Sint tres quantitates *AB*, *GD*, *E* et *AB* quantitas quantitate *GD* dato maior sit quam in proportione, similiterque quantitas *GD* quantitate *E* dato maior sit quam in proportione. Dico quod et *AB* quantitas quantitate *E* dato maior est quam in proportione.

Quoniam enim quantitas *GD* quantitate *E* dato maior est quam in proportione, auferatur data quantitas *GZ*. Reliqui ergo *ZD* ad *E* proportio est data. Rursum quoniam quantitas *AB* quantitate *GD* dato maior est quam in proportione, auferatur data quantitas *AI*. Reliqui ergo *IB* ad *GD* proportio est data. Eadem ei fiat quantitatis *IT* ad *GZ*. Proportio ergo quantitatis *IT* ad *GZ* data. Data autem *GZ*. Data ergo et *IT*. Estque et *AI* data. Totum ergo *TA* datum est. Et quoniam est sicut *IB* ad *GD* ita *IT* ad *GZ*, et reliqui *TB* ad reliquum *ZD* proportio est data. Et quantitatis *ZD* ad *E* proportio est data. Et quantitatis ergo *TB* ad *E* proportio est data. Et data *TA*. Quantitas ergo *BA* quantitate *E* dato maior est quam in proportione.

1 TB[1]....data[2] *illeg. D* / *post* data[2] *scr. B* est
2 date *om. B* / *post* AB[2] *add. mg. O* per XV
2–3 EZ....proportionem *illeg. D*
3 *de* vel *add. mg. D* per 15
5 XIX: 19[a] *B* / tres: 3 *B* / primaque: 1[a] que *B* / secunda: 2[a] *B*
6 pro[portione. . . . quoque *illeg. D* / secunda: 2[a] *B* / *post* quoque *scr. B* et / *post* maior *scr. B* est
7 *ante* prima *scr. B* et / tertia: 3[a] *B* / da[to. . . . quam *illeg. D* / erit *om. B*
8 tres: 3 *B*
9–10 qua[ntitas....dato *illeg. D*
11 E *corr. B ex* GD / maior....proportione *illeg. D* / est: sit *B*
12–13 propor[tione....data *illeg. D*
13 data *om. B*
14 qua[ntitas....GD *illeg. D*
15–16 ergo....data *illeg. D*
17 quantitatis *om. B* / ad....GZ *illeg. D* / autem *om. B* / *ante* Data[2] *scr. B* et
18 *ante* Totum *scr. et del. B* D / *de* Totum *add. mg. D* ex 3[a] / *ante* TA *scr. et del. B* DA dat
18–19 sicut....GZ *illeg. D*
19 *de* reliqui *add. mg. D* ex 19 5[ti]
20 ad....quantitatis *illeg. D* / est[1] *om. OP* / *post* ergo *add. mg. O* ex VIII
21 quantitate....maior *illeg. D* / *ante* E *scr. et del. B* E data

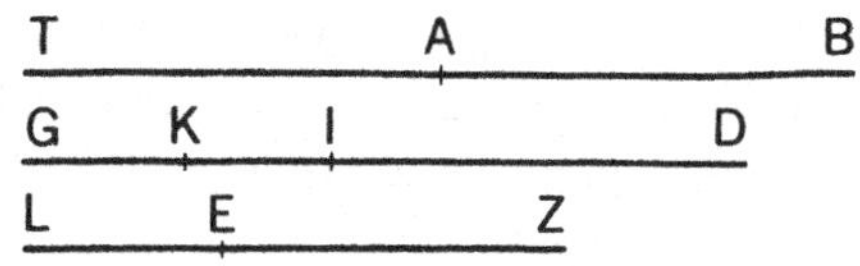

the ratio of *GK* to *LE* be made. Then *GK*:*LE* is given. But *GK* is given. Therefore *LE* is also given. The ratio of the whole *GD* to the whole *LZ* is given. *GD*:*TB* is also given. Therefore *TB*:*LZ* is also given. And from them [*TB*, *LZ*] let the given quantities *TA* and *LE* be taken away. Then either the quantities *AB* and *EZ* have a given ratio to each other or else the one is greater than the other by a given quantity as in ratio [*i.e.* or the excess of the one over a given quantity has a given ratio to the other].

XIX. If there are three quantities and if the first is greater than the second by a given quantity as in ratio and also the second is greater than the third by a given quantity as in ratio, the first is also greater than the third by a given quantity as in ratio.

Let the three quantities be *AB*, *GD* and *E*, and let *AB* be greater than *GD* by a given quantity as in ratio [*i.e.* let the excess of *AB* over a given quantity have a given ratio to *GD*]. And in similar way, *GD* is greater than *E* by a quantity as in ratio [*i.e.* the excess of *GD* over a given quantity has a given ratio to *E*]. I say that *AB* is also greater than *E* by a given quantity as in ratio [*i.e.* the excess of *AB* over a given quantity has a given ratio to *E*].

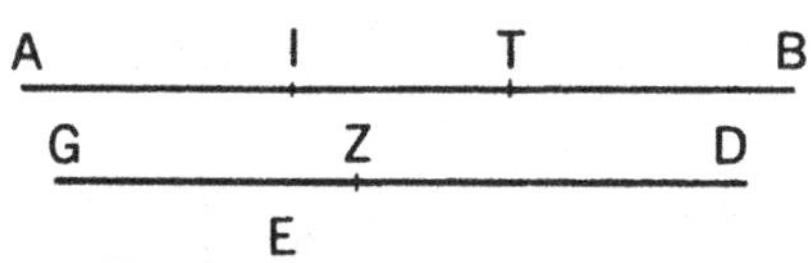

Since *GD* is greater than *E* by a given quantity as in ratio [*i.e.* since the excess of *GD* over a given quantity has a given ratio to *E*], let the given quantity *GZ* be taken away [from *GD*]. Then the ratio of the remainder *ZD* to *E* is given. Further, since *AB* is greater than *GD* by a given quantity as in ratio [*i.e.* since the excess of *AB* over a given quantity has a given ratio to *GD*], let the given quantity *AI* be taken away. Then the ratio of the remainder *IB* to *GD* is given. In the same way, let the ratio of *IT* to *GZ* be made. Therefore the ratio *IT*:*GZ* is given. But *GZ* is given. Therefore

Aliter

Sint tres quantitates *AB*, *G*, *D*, quantitasque *AB* quantitate *G* dato maior sit quam in proportione, quantitasque *G* quantitate *D* dato maior quam in proportione. Dico quod et quantitas *AB* quantitate *D* dato maior est quam in proportione.

Quoniam enim quantitas *AB* quantitate *G* dato maior est quam in proportione, auferatur data quantitas *AE*. Reliqui ergo *EB* ad *G* proportio data est. Et *G* quantitas quantitate *D* dato maior est quam in proportione. Et quantitas ergo *EB* quantitate *D* dato maior est quam in proportione. Auferatur ergo data quantitas *EZ*. Reliqui ergo *ZB* ad *D* proportio est data. Estque data *AZ*. Quantitas ergo *AB* quantitate *D* dato maior est quam in proportione.

(61r) /XX. Si fuerint due quantitates date, auferanturque ab eis quantitates adinvicem proportionem habentes datam, reliqua adinvicem vel proportionem habebunt datam vel alterum altero dato maius est quam in proportione.

Sint due quantitates date *AB*, *GD*, auferanturque ab eis quantitates *AE*, *GZ* proportionem habentes adinvicem datam. Dico quod *EB*, *ZD* adin-

2 tres: 3 *B* / G,D:GD *D*
2–3 quanti[tasque. . . .proportione *illeg. D*
3 dato maior *corr. B ex* maior dato
3–4 ma[ior. . . .quantitate *illeg. D*
4 est *om. B*
6–7 quantitas. . . .proportione *illeg. D*
7 auferatur *tr. B post* AE / *ante* quantitas *scr. B* et
7–8 ergo. . . .dato *illeg. D*
8 data est: est data *B* / dato: data *B*
9 D: B *B* / dato: data *B*
9–10 quan[titas. . . .Aufera]tur *illeg. D*
10 ergo[1] *om. B* / ZB:GB *B*
10–11 Re[liqui. . . .A]B *illeg. D*
13 XX: 20[a] *B* / Si. . . . eis *illeg. D* / due: 2 *B* / auferaturque: auferatur *P*
14 ad[invicem. . . . adin]vicem *illeg. D*
15–16 habe[bunt. . . . prop]ortione *illeg. D*
17 due: 2 *B*
17–18 quantita[tes. . . .adinvicem *illeg. D*
18 habentes *tr. B post* adinvicem
18–2(p.s.) Dico. . . .quam *illeg. D*

IT is given. And *AI* is also given. Therefore the whole *TA* is given. And since $IB{:}GD = IT{:}GZ$, the ratio of the remainder *TB* to the remainder *ZD* is also given. And the ratio $ZD{:}E$ is given. And, therefore, the ratio $TB{:}E$ is given. And *TA* is given. Therefore *BA* is greater than *E* by a given quantity in ratio [*i.e.* the excess of *BA* over a given quantity has a given ratio to *E*].

Another proof

Let the three quantities be *AB*, *G* and *D*, and let *AB* be greater than *G* by a given quantity as in ratio [*i.e.* the excess of *AB* over a given quantity has a given ratio to *G*], and let *G* be greater than *D* by a given quantity as in ratio [*i.e.* let the excess of *G* over a given quantity have a given ratio to *D*]. I say that *AB* is also greater than *D* by a given quantity as in ratio [*i.e.* the excess of *AB* over a given quantity has a given ratio to *D*].

Since *AB* is greater than *G* by a given quantity as in ratio [*i.e.* since the excess of *AB* over a given quantity has a given ratio to *G*], let a given quantity *AE* be taken away. The ratio of the remainder *EB* to *G* is given. And

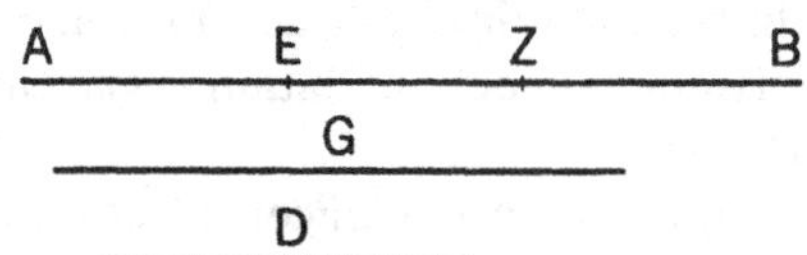

G is greater than *D* by a given quantity as in ratio [*i.e.* the excess of *G* over a given quantity has a given ratio to *D*]. Therefore *EB* is greater than *D* by a given quantity as in ratio [*i.e.* the excess of *EB* over a given quantity has a given ratio *D*]. Therefore, let the given quantity *EZ* be taken away. Then the ratio of the remainder *ZB* to *D* is given. And *AZ* is given. Therefore *AB* is greater than *D* by a given quantity as in ratio [*i.e.* the excess of *AB* over a given quantity has a given ratio to *D*].

XX. If there are two given quantities and from them there are taken quantities having a given ratio to each other, either the remaining quantities will have a given ratio to each other or else the one will be greater than the other by a given quantity as in ratio.

Let the two given quantities be *AB* and *GD* and from then let there be taken the quantities *AE* and *GZ*, having a given ratio to each other. I say

vicem vel proportionem habent datam vel alterum altero dato maius est quam in proportione.

Quoniam enim utrumque *AB*, *GD* datum est, proportio ergo quantitatis *AB* ad *GD* data. Et si eadem est quantitatis *AE* ad *GZ*, erit quoque reliqui *EB* ad reliquum *ZD* proportio data.

Non sit autem eadem, fiatque sicut *EA* ad *GZ* ita *AI* ad *GD*. Proportio autem quantitatis *AE* ad *GZ* data. Proportio ergo et quantitatis *AI* ad *GD* data. Datum autem *GD*. Datum ergo et *AI*. Est autem et *AB* data. Et reliquum ergo *IB* datum est. Et quoniam est sicut *AE* ad *GZ* ita *AI* ad *GD*, et reliqui *IE* ad reliquum *ZD* proportio est data. Datum autem *IB*. Quantitas ergo *EB* quantitate *ZD* dato maior est quam in proportione.

XXI. Si fuerint due quantitates date, apponanturque eis quantitates adinvicem proportionem habentes datam, tota adinvicem vel proportionem habebunt datam vel alterum altero dato maius est quam in proportione.

Sint due quantitates date *AB*, *GD*, apponanturque eis quantitates *AE*, *GZ* proportionem habentes adinvicem datam. Dico quod tota *EB*, *ZD* adinvicem vel proportionem habebunt datam vel aterum altero dato maius est quam in proportione.

Quoniam enim utraque data est videlicet *AB* et *GD*, proportio ergo quantitatis *AB* ad *GD* data est. Et si est eadem ei que est quantitatis *EA* ad *GZ*, et totius *EB* ad totum *ZD* proportio data.

Si vero non, fiat sicut *AE* ad *GZ* ita *IA* ad *GD*. Proportio ergo quantitatis *IA* ad *GD* data. Data autem *GD*. Data ergo et *IA*. Estque et *AB* datum. Et reliquum ergo *IB* datum est. Et quoniam est sicut *EA* ad *GZ*

3 datum: data *B*
3–4 Quoniam. . . .est *illeg. D*
4 AE:EA *B* / erit quoque reliqui: reliqui ergo erit *B*
4–6 ad[2]. . . .eadem *illeg. D*
6–7 EA. . . .quanti]tatis[2] *illeg. D*
7–9 GD. . . .IB *illeg. D*
8 AI:IA *B*
9 *de* quo- *add. mg. D* ex 3 / *ante* ita *scr. B* et
9–10 quo[niam. . . .data *illeg. D*
10–11 autem. . . .proportione *illeg. D*
12 XXI: 21[a] *B* / Si. . . . quan]titates *illeg. D* / due: 2 *B*
13–14 datam. . . . da]tam *illeg. D*
14 dato *om. BO*
14–16 alt[erum. . . . eis *illeg. D*
16 due: 2 *B* / date *om. B*
17 adinvicem *tr. B post* datam
17–18 haben[tes. . . .propor]tionem *illeg. D*
18 dato *om. B*
18–20 d[ato. . . .est *illeg. D*
21 ad. . . .EA *illeg. D* / est[1] *om. B* / quantitatis: quantitas *P*
22 totius: tocius *BP* / *de* totius *add. mg. D* ex 12, 5 euclidis
23 vero. . . .GD *illeg. D*
24 IA[1]: I *D*
24–25 Da[ta[2]. . . .reliquum *illeg. D*

that *EB* and *ZD* have either a given ratio to each other or else the one is greater than the other by a given quantity as in ratio [*i.e.* the excess of the one over a given quantity has a given ratio to the other].

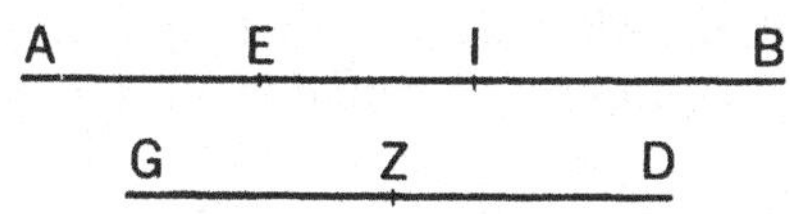

Since each of *AB* and *GD* is given, the ratio *AB*:*GD* is given. And if it is the same as *AE*:*GZ*, the ratio of the remainder *EB* to the remainder *ZD* will also be given.

Let it be not the same as it [*i.e.* as *AE*:*GZ*] but let *EA*:*GZ* = *AI*:*GD*. Now *AE*:*GZ* is given. Therefore the ratio *AI*:*GD* is also given. But *GD* is given. Therefore *AI* is also given. Now *AB* is also given. Then the remainder *IB* is also given. Since *AE*:*GZ* = *AI*:*GD*, the ratio of the remainder *IE* to the remainder *ZD* is also given. Now *IB* is given. Therefore *EB* is greater than *ZD* by a given quantity as in ratio [*i.e.* the excess of *EB* over a given quantity has a given ratio to *ZD*].

XXI. If there are two given quantities and to them are added other quantities having a given ratio to each other, either the wholes will have a given ratio to each other or else the one is greater than the other by a given quantity as in ratio.

Let the two given quantities be *AB* and *GD* and to them let there be added the quantities *AE* and *GZ*, having a given ratio to each other. I say that the wholes *EB* and *ZD* have a given ratio to each other or else the one is greater than the other by a given quantity as in ratio [*i.e.* the excess of the one over a given quantity has a given ratio to the other].

Since both quantities, that is, *AB* and *GD*, are given, the ratio *AB*:*GD* is given. And if it is the same ratio as *EA*:*GZ*, the ratio of the whole *EB* to the whole *ZD* is given.

But if it is not the same, let it be that *AE*:*GZ* = *IA*:*GD*. Then the ratio *IA*:*GD* is given. Now *GD* is given. Therefore *IA* is also given. And *AB* is also given. Therefore the remainder *IB* is also given. Since *EA*:*GZ* = *AI*:*GD*, the ratio of the whole *EI* to the whole *ZD* is also given. But *IB*

ita *AI* ad *GD*, et totius *EI* ad totum *ZD* proportio est data. Data autem *IB*. Quantitas ergo *EB* quantitate *ZD* dato maior est quam in proportione.

XXII. Si due quantitates ad aliquam quantitatem proportionem habent datam, et coutrumque ad eandem proportionem habet datam.

Due enim quantitates *AB*, *BG* ad aliquam quantitatem *D* proportionem habeant datam. Dico quod et coutrumque *AG* ad *D* proportionem habet datam.

Quoniam enim utrumque *AB* et *BG* ad *D* proportionem habet datam, proportio ergo et quantitatis *AB* ad *BG* data. Et componenti quantitatis *AG* ad *GB* proportio est data. Et quantitatis *GB* ad *D* proportio est data. Et quantitatis ergo *AG* ad *D* proportio est data.

XXIII. Si totum ad totum proportionem habet datam, partesque ad partes habent proportiones datas, non autem eadem, et omnia ad omnia proportiones habebunt datas.

Habeat enim totum *AB* ad totum *GD* proportionem datam, habeantque et *AE* et *EB* partes ad *GZ* et *ZD* partes proportiones datas, non autem easdem. Dico quod et omnia ad omnia proportiones habebunt datas.

Quoniam enim proportio quantitatis *AE* ad *GZ* data, eadem ei fiat quantitatis *AB* ad *GI*. Proportio ergo et quantitatis *AB* ad *GI* data. Eritque reliqui *EB* ad reliquum *ZI* proportio data. Quantitatis quoque *EB* ad *ZD*

1 ita. . . .et *om. P* / et. . . .data *illeg. D*
2 *ante* quantitate *scr. et del. B* AG ad D proportio data / dato: data *B* / quam in proportione *illeg. D*
3 XXII: 22^{a} *B*
3–4 propor[tionem . . . co]utrumque *illeg. D*
5 quantitates. . . .ali]quam *illeg. D*
5–6 quantitatem. . . .habeant *illeg. D*
6 coutrumque: utrumque *P*
8 utrumque. . . .D *illeg. D*
8–9 utrumque. . . .ergo *om. P*
9 *post* ad *add. mg. O* ex VIII / *de* componen- *add. mg. D* ex 6^{a} / *supra* componenti *add. O* ex VI
9–10 component[ti. . . .GB^{1} *illeg. D*
10 *post* Et *scr. et del. D* componenti / *ante* GB^{2} *scr. et del. D* A / est^{2} *om. B*
11 Et *om. B* / Et. . . .propor]tio *illeg. D* / est *om. BP*
12 XXIII: 23^{a} *B*
13 habent *tr. B post* proportiones / proportiones. . . . non *illeg. D*
14–15 datas. . . . totum^{2} *illeg. D*
16 et^{1}: etiam *B*
16–17 datas. . . .Dico *illeg. D*
17 et *om. B* / proportiones *om. B*
18 quantita[tis. . . .ea]dem *illeg. D*
19 Proportio. . . .GI *om. B*
19–20 GI^{2}. . . .EB^{1} *illeg. D*
20 reliqui: reliquum *B*

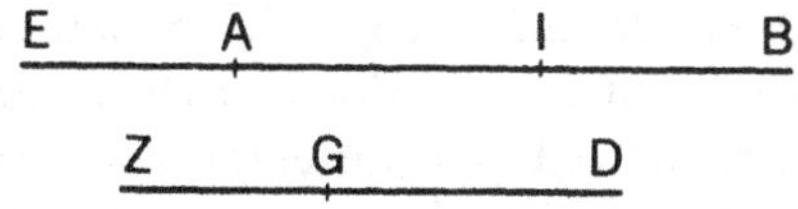

is given. Therefore *EB* is greater than *ZD* by a given quantity as in ratio [*i.e.* the excess of *EB* over a given quantity has a given ratio to *ZD*].

XXII. If two quantities have a given ratio to some other quantity, their sum has a given ratio to the same quantity.

Let two quantities *AB* and *BG* have a given ratio to some other quantity *D*. I say that their sum *AG* has a given ratio to *D*.

Since each of *AB* and *BG* has a given ratio to *B*, the ratio *AB*:*BG* is also given. And, by composing, the ratio *AG*:*GB* is given. And the ratio *GB*:*D* is given. Therefore the ratio *AG*:*D* is given.

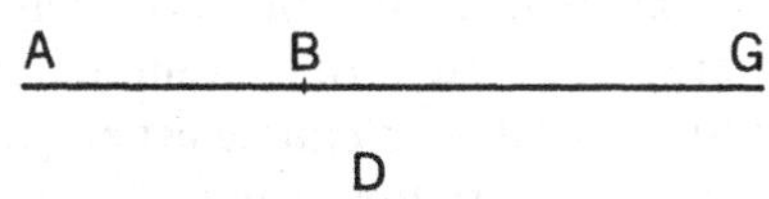

XXIII. If the whole has a given ratio to the whole and if the parts have given ratios to the parts, but not the same, all [the quantities] will have given ratios to all [the quantities].

Let the whole *AB* have a given ratio to the whole *GD* and let the parts *AE* and *EB* have given ratios to the parts *GZ* and *ZD*, but not the same. I say that all [the quantities *AB*, *AE*, *EB*] will also have given ratios to all [the quantities *GD*, *GZ*, *ZD*].

Since the ratio *AE*:*GZ* is given, let the same as it be made of *AB* to *GI*. Then the ratio *AB*:*GI* is also given. And the ratio of the remainder *EB* to the remainder *ZI* will be given. The ratio *EB*: *ZD* is also given. And,

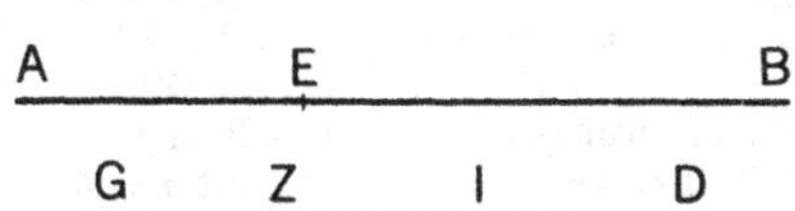

proportio est data. Et quantitatis ergo *ZD* ad *ZI* proportio est data. Et convertenti quantitatis *ZD* ad *DI* proportio est data. Et quoniam proportio est quantitatis *BA* ad utrumque videlicet *DG* et *GI*, et quantitatis ergo *DG* ad *GI* proportio est data. Et convertenti quantitatis *GD* ad *DI* proportio est data. Sed quantitatis *ID* ad *DZ* proportio est data. Et quantitatis ergo *GD* ad *DZ* proportio est data. Quare et quantitatis *GZ* ad *ZD* proportio est data. Sed et quantitatis *GZ* ad *AE* proportio est data. Quantitatis autem *ZD* ad *BE* proportio est data. Quare omnium ad omnia proportio est data.

(61v) /XXIIII. Si tres recte proportionales fuerint, primaque ad tertiam proportionem habeat datam, et ad secundam proportionem habebit datam.

Sint tres recte proportionales *A*, *B*, *G* sicut *A* ad *B* ita *B* ad *G*, rectaque *A* ad *G* proportionem habeat datam. Dico quod et ad *B* proportionem habet datam.

Ponatur enim data *D*. Et quoniam proportio est quantitatis *A* ad *G* data, eadem ei fiat quantiatis *D* ad *Z*. Proportio ergo quantitatis *Z* ad *D* data. Data autem *D*. Data ergo et *Z*. Sumatur autem quantitatum *D*, *Z* media proportionalis *E*. Quod ergo sub *D Z* equale est ei quod ab *E*. Datum autem quod sub *D Z*, data enim utraque earum. Datum ergo et quod ab *E*. Data ergo et *E*. Data autem est et *D*. Proportio ergo quantitatis *D* ad *E*

1 data. . . .ad *illeg. D*
2 data. . . . quanti]tatis *illeg. D* / quoniam: quia *B*
2–3 proportio *tr. B post* est
3 quantitatis[1] *om. B* / *supra* quantitatis[2] *add. O* ex VIII
4 GI. . . .convertenti *illeg. D* / GD: DG *B* / DI:GI *B*
5–6 Sed. . . .data *om. P* / est[2]. . . .GD *illeg. D*
6 *post* GD *add. mg. O* ex VIII
6–7 proportio[2]. . . .AE *illeg. D*
7 *supra* est data[1] *add. O* ex V et VIII
8 omnia: totum *B*
8–9 B[E. . . .data *illeg. D*
9 proportio est data: data est proportio data *B*
10 XXIIII: 24ª *B* / tres: 3 *B* / primaque: 1ª que *B* / tertiam: 3ª *B correxi ex* terciam *OP hic et ubique*
10–11 f[uerint. . . . habeat *illeg. D*
11 secundam: 2ª *B*
11–14 proportionem[2]. . . . A *illeg. D*
13 *post* tres *scr. B* sicut
14 proportionem[1]: proportio *B* / habeat: habet *P et om. B* / datam: data *B*
14–16 Dico. . . .proportio *illeg. D*
15 habet: habeat *P*
16 quoniam: quia *B*
17 eadem. . . .D[2] *illeg. D* / eadem *tr. P post* ei / quantitatis[1] *om. B* / *ante* quantitatis[2] *scr. P* quantitatis D ad Z
18 et: est *B* / quantitatum: quantitas *B* / Z[2]: et *B*
18–19 Z[1]. . . .ergo *illeg. D*
19–20 E[2]. . . .ab *illeg. D*
21 est *om. B*
21–1(p.s.) autem. . . .quod[1] *illeg. D*

therefore, the ratio $ZD:ZI$ is given. And, by converting, the ratio $ZD:DI$ is given. Since the ratios $BA:DG$ and $BA:GI$ are given, the ratio $DG:GI$ is also given. And, by converting, the ratio $GD:DI$ is given. But the ratio $ID:DZ$ is given. Therefore the ratio $GD:DZ$ is given. Hence the ratio $GZ:ZD$ is also given. But the ratio $GZ:AE$ is given. How the ratio $ZD:BE$ is given. Hence the ratio of all [the quantities AB, AE, EB] to all [the quantities GD, GZ, ZD] is given.

XXIV. If there are three straight lines and the first has a given ratio to the third, it will also have a given ratio to the second.

Let there be three proportional straight lines A, B and G, that is, $A:B = B:G$, and let a straight line A have a given ratio to G. I say that it also has a given ratio to B.

Let D be posited as given. Since the ratio $A:G$ is given, let the same as it be made of D to Z. Then the ratio $Z:D$ is given. But D is given. Therefore Z is also given. Now let there be taken a mean proportional E of the quantities D and Z. Then, the product of D and Z is equal to the square of E [*i.e.* $D \times Z = E^2$]. Now $D \times Z$ is given, for each of them [D, Z] is given. Therefore E^2 is given. Therefore E is also given. But D is also given. Therefore the ratio $D:E$ is given. Since* $A:G = A^2:(A \times G)$ and also $D:Z = D^2:(D \times Z)$, therefore $A^2:(A \times G) = D^2:(D \times Z)$. But $A \times G = B^2$, for A, B and G are proportional. And $D \times Z = E^2$. Therefore $A^2:B^2 = D^2:E^2$. Therefore $A:B = D:E$. Now the ratio $D:E$ is given. Therefore $A:B$ is also given.

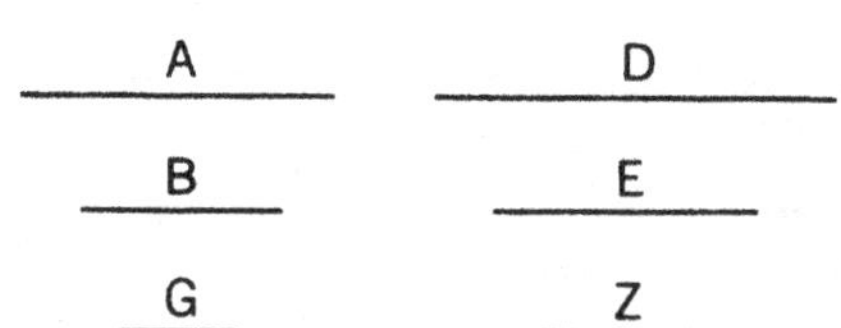

* Greek text adds here after *Since* "$A:G = D:Z$, but indeed", apparantly not in the translator's copy; or at least he did not translate it into Latin.

data. Et quoniam est sicut *A* ad *G* ita quod ab *A* ad illud quod sub *A G*, sicut etiam *D* ad *Z* ita quod ab *D* ad illud quod sub *D Z*, sicut ergo quod ab *A* ad illud quod sub *A G* ita quod ab *D* ad illud quod sub *D Z*. Sed quod sub *A Z* equale est ei quod ab *B*, proportionales enim sunt *A*, *B*, *G*. Et quod sub *D Z* equale est ei quod ab *E*. Sicut ergo quod ab *A* ad illud quod ab *B* ita quod ab *D* ad illud quod ab *E*. Et sicut ergo *A* ad *B* ita *D* ad *E*. Proportio autem linee *D* ad *E* data. Proportio ergo et linee *A* ad *B* data.

Aliter de eodem

Quoniam proportio est linee *A* ad *G* data sicutque *A* ad *G* ita quod ab *A* ad illud quod sub *A G*, proportio ergo et eius quod ab *A* ad illud quod sub *A G* data. At quod sub *A G* equale est ei quod ab *B*. Proportio ergo eius quod ab *A* ad illud quod ab *B* data. Quare et linee *A* ad *B* proportio est data.

XXV. Si due linee positione date se adinvicem secent, punctus quoque in quo se invicem secant, datus est positione.

Due enim linee positione date *AB*, *GD* se adinvicem secent ad punctum *E*. Dico quod datus est punctus *E*.

Si enim non, excidet *E* punctus. Excidet ergo et unius harum *AB*, *GD* positio. At non excidit. Datus ergo punctus *E*.

XXVI. Si recte linee termini dati fuerint positione, data erit recta positione et magnitudine.

1 quoniam: quia *B* / *post* ita *add. mg. O* per primam VI^{ti} elementorum / ab *bis P*
1–2 A G. . . .D Z *illeg. D*
2 etiam: et *B* / *ante* sicut^{2} *scr. et del. O* sed quod sub A G
3–4 ad^{1}. . . .ab *illeg. D*
4 Z: G *B*
4–6 enim. . . .B^{1} *illeg. D*
6–7 ad^{1}. . . .Proportio^{2} *illeg. D*
7 autem: ergo *B* / *ante* ergo *scr. mg. D* data ad quod sub A G / et *om. B*
7–8 linee^{2}. . . .data *illeg. D*
9–14 Aliter. . . .data *om. D*
10 proportio *tr. B post* est
10–13 ab *correxi ex* sub *BOP*
13 A. . . .ab *om. B*
15 XXV: 5^{a} *B*
15–20 Si. . . .E *illeg. D*
16 secant: secent *B*
17 adinvicem: invicem *B*
19 E *bis P*
20 excidit: excideret *B* / *supra* excidit *add. O* ex diffinitione / E: est *O*
21 XXVI: 26^{a} *B* / dati *tr. P post* fuerint / fuerint *tr. B ante* termini / Si. . . . dat]a *illeg. D*
22–1(p.s.) et. . . .positione *illeg. D*

Another proof for the same

Since the ratio $A:G$ is given and $A:G = A^2:(A \times G)$, the ratio $A^2:(A \times G)$ is also given. But $(A \times G) = B^2$. Therefore the ratio $A^2:B^2$ is given. Hence the ratio $A:B$ is also given.*

XXV. If two lines given in position intersect, the point at which they intersect each other is also given in position.

Let two lines, AB and GD, given in position intersect each other at the point E. I say that the point E is given.

If the point E is not given in position, it will change its position. Therefore one of the lines AB or GD will change its position. But it does not change position. Therefore the point E is given.

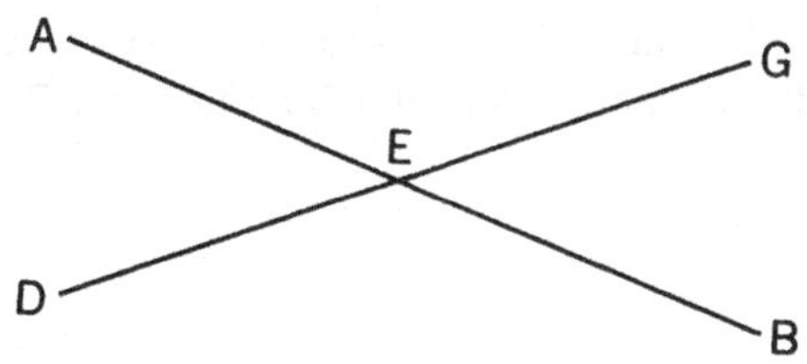

XXVI. If the extremities of a straight line will have been given in position, the straight line will be given in position and in magnitude.

* Greek adds (in margin of Vat. Gr. 190 with sign of omission) an additional phrase "for we have found in the squares of A and B, straight lines which are equal to these lines." This is another indication that the translator had a good manuscript. See Menge, p. 192.

Recte enim linee termini *A*, *B* dati sint positione. Dico quod data est recta *AB* positione et magnitudine.

Si enim manente *A* excidet *AB* recte vel positio vel magnitudo, excidet et *B* punctus. Atqui non excidit. Data ergo recta *AB* positione et magnitudine.

XXVII. Si recte linee positione et magnitudine date unus terminus datus fuerit, alter quoque datus erit.

Recte enim linee positione et magnitudine date *AB* unus terminus *A* sit datus. Dico quod et *B* terminus datus est.

Si enim manente *A* puncto excidet *B* punctus, excidet ergo et *AB* recte vel positio vel magnitudo. Atqui non excidit. Datus est ergo *B* punctus.

Aliter

Centro *A* spatio vero *AB* periferia describatur *GBD*. Positione est <data> *GBD*. Positione quoque et *AB* recta data. Datus ergo *B* punctus.

1 enim *om. O* / dati: data *B* / sint: sunt *B*
2 positione et magnit]udine *illeg. D*
3 manente....vel[2] *illeg. D* / excidet: excidat *B* / *de* magnitudo *add. mg. O* Ex expositione diffinitionis magnitudinis date in magnitudine superius proposita, sequitur quod si aliquid sit datum magnitudine, eius magnitudo fixa est super esse unum scilicet nec crescens nec decrescens. Unde linea ducta a centro terre in centrum solis non est linea data magnitudine. Cuius igitur magnitudo excidit, datum non est magnitudine et econverso et quod est datum magnitudine, eius magnitudo non excidit. Sine horum intellectu nichil arguitur hic nec posterius.
3–5 magni[tudo....magni]tudine *illeg. D*
4 excidit: exciditi *B* / *supra* excidit *add. O* ex diffinitione / recta *om. B*
6 XXVII: 27[a] *B* / Si .... magnitu]dine *illeg. D*
7 datus .... erit *illeg. D* / alter: reliquus *B*
9 *post* terminus *scr. B* eius
10 Si....excidet[1] *illeg. D* / puncto *om. B* / excidet[1] : excidit *O*
11 positio: positione *B* / magnitudo: magnitudine *B* / Atqui....Datus *illeg. D* / excidit: excidet *B*
13 Positione: Positioneque *B* / *de* Positione *add. mg. O* quia eius centrum est datum / *ante* est *scr. O* quoque
13–14 GBD....Positione *illeg. D* / est....quoque *om. B*
14 AB:B *D* / *de* Datus *add. mg. D* quia eius centrum datum est / *post* punctus *add. mg. P* per XXV

Let the extremities *A* and *B* of a straight line be given in position. I say that the straight line *AB* is given in position and magnitude.

If, with the point *A* remaining in its place, the position and magnitude of the straight line *AB* changes, the point *B* also changes its position. Now [in fact] it does not change its position. Therefore the straight line *AB* is given in position and magnitude.

XXVII. If one extremity of a straight line given in position and magnitude will have been given, the other extremity will be also given.

Let one extremity of the straight line *AB* given in position and magnitude, be given. I say that the other extremity *B* is also given.

If, with point *A* remaining in its position, point *B* changes its position, the position or magnitude of the straight line *AB* will also change. But [by supposition] neither [the position nor the magnitude of *AB*] changes. Therefore the point *B* is given.

Another proof

With the center *A* and the [radial] distance *AB*, let the circumference *GBD* be described. [Then] *GBD* is given in position. The straight line *AB* is also given in position. Therefore the point *B* is given.

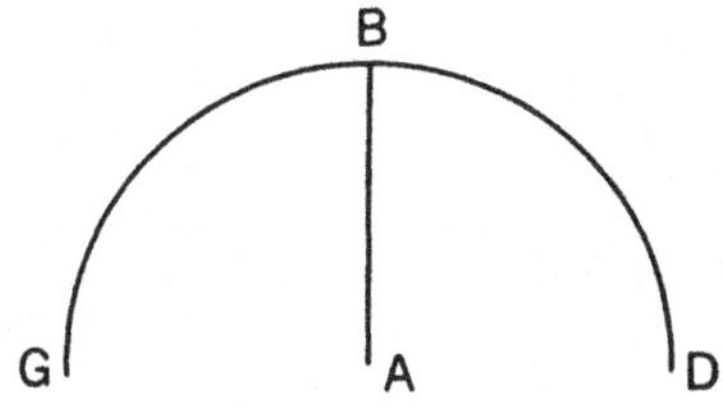

XXVIII. Si per datum punctum secundum positione datam rectam recta linea protrahatur, data erit et protracta positione.

Per datum enim punctum *A* secundum positione datam rectam *BG* recta linea protrahatur *DAE*. Dico quod data est *DAE* positione. Si enim minime, manente *A* puncto excidet recte *DAE* positio. Permanente *BG* equidistante excidat et esto *ZAI*. Equidistans ergo est recta *GB* recte *ZAI*. Atqui *BG* recta recte *DAE* equidistans est. Et recta ergo *DAE* recte *ZAI* equidistans est. Sed concidunt, quod est inconveniens. Non ergo excidet recte *DAE* positio. Positione ergo est <data> *DAE*.

XXIX. Si a positione data recta et in ea dato puncto recta linea protrahatur datum faciens angulum, data est protracta positione.

A positione enim data recta *AB*, atque in ea dato puncto *G* recta protrahatur *GD* datum faciens angulum *BGD*. Dico quod positione est data *GD*.
(62r) Si enim minime, manente *G* puncto excidet recte/*GD* positio observans anguli *BGD* magnitudinem. Excidat et esto *GE*. Equalis est ergo angulis *DGB* angulo *EGB* maior minori, quod est inconveniens. Non ergo excidet recte *DG* positio. Positione ergo est *DG*.

1 XXVIII: 28[a] *B* / Si . . . . punctum *illeg. D* / positione: positionem *B* / rectam *om. D*
2 erit . . . . positione *illeg. D* / protracta *tr. B post* positione
3 positione: positionem *B*
4 protra[hatur DAE *illeg. D*
5 puncto. . . .DAE *illeg. D* / recte: recta *B* / *post* positio *add. mg. O* ex diffinitione
5–6 *de* Permanente *add. mg. DO* ex propositione et diffinitione
6 *post* ergo *scr. P* E
6–7 ergo. . . .ZAI *illeg. D* / est. . . . est *om. P*
7 GB:BG *B* / ZAI:ZA *B* / Atqui: Sed *B* / est *om. B* / *de* ergo *add. mg. D* ex propositione elementorum
8 equidistans. . . .concidunt *illeg. D*
10 XXIX: 29[a] *B* / Si . . . . data *illeg. D*
11 da[tum . . . . angulum *illeg. D*
12 *ante* recta[2] *scr. B* et
12–13 pro[trahatur. . . .angulum *illeg. D*
13 GD[1]:DG *B* / BGD:BD *B*
14 minime: non *B* / *de* manente *add. mg. D* ex diffinitione
14–15 excidet. . . .anguli *illeg. D*
15 magnitudinem: magnitidine *B* / ergo *om. B*
16 DGB:GDB *B* / an[gulo. . . .est *illeg. D*
17 recte *tr. B ante* positio / DG[1]:GD *B* BG *O* / DG[2]:GD *B*

XXVIII. If through a given point there is drawn a straight line parallel to a straight line given in position, the straight line so drawn will also be given in position.

Through the given point *A* let there be drawn a straight line *DAE* parallel to the straight line *BG* given in position. I say that *DAE* is given in position.

If it is not so given, with the point *A* remaining in its place, the position of the straight line *DAE* will change. With it always remaining parallel to *BG*, let it change and be *ZAI*. Then the straight line *GB* is parallel to the straight line *ZAI*. But the straight line BG is parallel to the straight line *DAE*. Therefore the straight line *DAE* is also parallel to the straight line *ZAI*. But they must concur with each other, which is absurd. Therefore the position of the straight line *DAE* does not change. Therefore *DAE* is given in position.

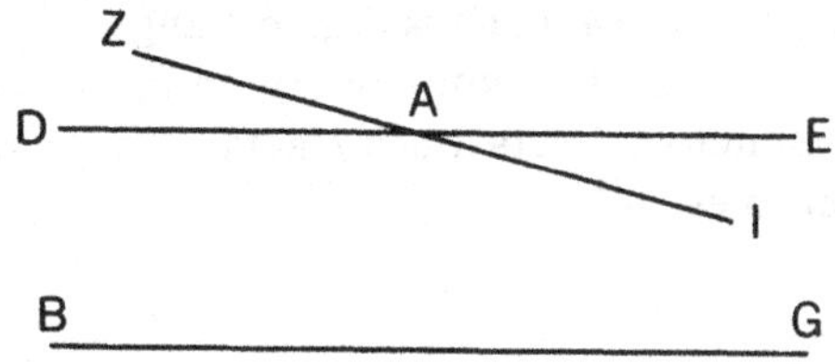

XXIX. If from a straight line given in position and from a point given therein there is drawn a straight line making a given angle, the line so drawn is given in position.

From the straight line *AB* given in position and from the point *G* given in it, let there be drawn a straight line *GD* making a given angle *BGD*. I say that *GD* is given in position.

If it is not so given, with the point *G* remaining in its place, the position of the line *GD* will change maintaining the magnitude of the angle *BGD*. Let its position change; and let it be *GE*. Then the angle *DGB* is equal to the angle *EGB*, the greater to the lesser, which is absurd. Therefore the position of the line *DG* will not change. Therefore *DG* is given in position.

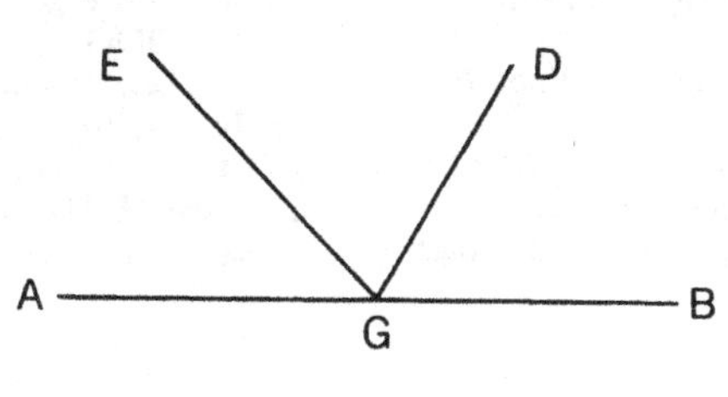

XXX. Si a dato puncto super positione datam rectam recta linea protrahatur datum faciens angulum, data est ea protracta positione.

A dato enim puncto *A* super positione datam rectam *BG* recta linea protrahatur *AD* datum faciens angulum *ADG*. Dico quod positione data est *AD*.

Si enim minime, manente *A* excidet recte *AD* positio observans anguli *ADG* mangitudinem. Excidat sitque *AZ*. Equalis est ergo angulus *ADG* angulo *AZG* maior minori, quod est impossibile. Non ergo excidet *AD*. Positione ergo est *AD*.

Aliter

Protrahatur per *A* punctum recte *BDG* equidistans recta *EAZ*. Quoniam ergo per datum punctum *A* secundum positione datam *BDG* recta linea tracta est *EAZ*, positione ergo est *EAZ*. Et quoniam equidistans est *EAZ* recte *BDG* eisque incidit *DA*, equalis ergo est angulus *EAD* angulo *ADG*. Datus est ergo angulus *EAD*. Quoniam ergo a positione data recta *EAZ* atque in ea dato puncto *A* recta linea tracta est datum faciens angulum *EAD*, positione ergo est *AD*.

1 XXX: 30[a] *B* / Si . . . punc]to *illeg. D* / a *om. P*
2 da[tum faciens angulum *illeg. D* / est *tr. P post* ea
3 A dato: D dato *B* / BG:GB *B*
3–4 linea. . . .ADG *illeg. D*
4 quod *om. B*
6 minime: non *B* / *de* ma- *add. mg. D* ex diffinitione
6–7 ma[nente. . . .magnitudinem *illeg. D*
7 magnitudinem: magnitudine *B* / ergo *om. B*
7–8 an[gulus. . . .impossibil]e *illeg. D*
10–11 Aliter. . . .recta *illeg. D*
11 BDG: BGD *D* / *de* Quoniam *add. mg. D* per 28
12 secundum *bis P* / positione: positionem *B* / datam *tr. B post* secundum / *post* recta *add. mg. O* per XXVIII
12–13 positione. . . .tracta *illeg. D*
13 EAZ[1]:ZAZ *B* / positione. . . .EAZ[2] *om. B*
13–14 Et. . . .est *om. D*
14 equalis *tr. B post* ADG / ergo est: est ergo *BP*
14–15 BDG. . . .es]t *illeg. D*
15 *de* ergo[2] *add. mg. D* ex premissa
15–16 positione. . . .trac]ta *illeg. D*
16 tracta *tr. B post* est
17 *supra* EAD *add. O* ex premissa
17–2(p.s.) EAD. . . .datu]s *illeg. D*

XXX. If from a given point to a straight line given in position there is drawn a straight line making a given angle, the line so drawn is given in position.

From a given point *A* to the straight line *BG* given in position let there be drawn the straight line *AD* making a given angle *ADG*. I say that *AD* is given in position.

If it is not so given, with *A* remaining in its position, the position of the straight line *AD* will change maintaining the magnitude of the angle *ADG*. Let it change its position and be *AZ*. Therefore the angle *ADG* is equal to the angle *AZG*, the greater to the less, which is impossible. Therefore *AD* will not change its position. Therefore *AD* is given in position.

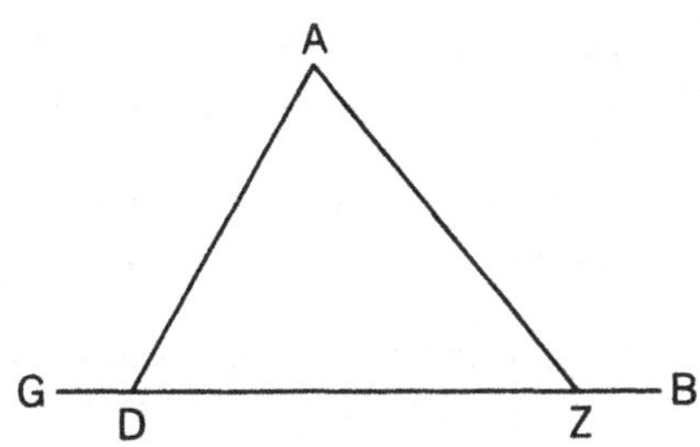

Another proof

Through the point *A* let there be drawn the straight line *EAZ* parallel to the straight line *BDG*. And since through the given point *A* and parallel to the straight line *BDG* given in position there is drawn the straight line *EAZ*, *EAZ* is given in position. Since *EAZ* is parallel to the straight line *BDG* and *DA* intersects them, the angle *EAD* is therefore equal to the angle *ADG*. Hence the angle *EAD* is given. Since from the straight line *EAZ* given in position and from the given point *A* in it there is drawn the straight line *AD* making a given angle *EAD*, hence *AD* is given in position.

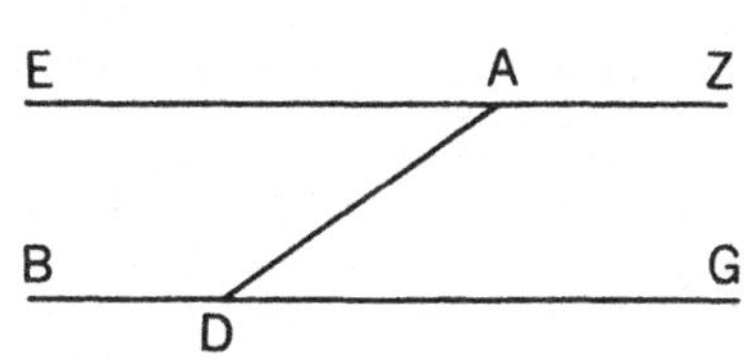

Aliter

Sumatur super *BG* datus punctus *E* et ab *E* puncto recte *AD* equidistans trahatur *EZ*. Et quoniam equidistans est recta *ZE* recte *AD* eisque incidit *BED*, equalis ergo est angulus *ZED* angulo *ADG*. Datus est ergo et *ZED*. Quoniam ergo a positione data recta *BG* et in ea dato puncto *E* recta linea tracta est *EZ* datum faciens angulum *ZEG*, positione ergo est *EZ*. Quoniam ergo a dato puncto secundum positione datam rectam *ZE* recta linea tracta est *AD*, positione ergo est *AD*.

Aliter

Recipiatur super *BG* forte punctus *E* et coniungatur *E A*. Quoniam datus est *A* punctus, positione ergo est *AE*. Positione quoque et *BG*. Datus ergo est angulus *AED*. Estque et angulus *ADE* datus. Et reliquus ergo *EAD* datus est. Quoniam ergo a positione data recta *EA* et in ea puncto *A* recta linea tracta est *AD* datum faciens angulum *EAD*, positione ergo est *AD*.

XXXI. Si a dato puncto super positione datam rectam linea recta ducatur data magnitudine, data est et positione.

A dato enim puncto *A* super positione datam rectam *BG* recta linea trahatur *DA* data magnitudine. Dico quod et positione data est.

2–3 recte. . . .recta *illeg. D*
4 ergo[1] *tr. B post* est[1] / ZED[2]:ZEB *B* / *post* ZED[2] *scr. B* angulus
4–5 BED. . . .positio]ne *illeg. D*
5–6 ea. . . .positio]ne *illeg. D*
6 EZ[2]:AD *B*
7 *post* a *scr. B* positione / puncto *om. B* / positione: positionem *B* / *post* rectam *add. mg. O* per antepremissam / recta: reliqua *B*
7–8 ergo. . . .AD[1] *illeg. D*
8 *post* AD[2] *add. mg. D* ex antepremissa
9–11 Aliter. . . .pun]ctus *illeg. D*
11 *supra* AE *add. O* ex XXVI / BG:DG *O* / ergo[2] *tr. B post* est *ln. 12*
11–12 est[2]. . . .reliqu]us *illeg. D*
12 et: etiam *B* / *supra* ADE datus *add. O* ex propositione / EAD:EAE *B*
13 est *om. BP*
13–14 datus. . . .est[1] *illeg. D*
14 *post* faciens *add .mg. O* ex XXIX / fa[ciens. . . .AD *illeg. D* / AD:EAD *B*
15 XXXI: 31[a] *B*
15–16 Si . . . . duc]atur *illeg. D*
16–17 da[ta[1]. . . . re]ctam *illeg. D*
17–18 recta. . . .magni]tudine *illeg. D*
18–2(p.s.) Dico. . . .circulus *illeg. D*

Another proof

On *BG* let there be taken the given point *E* and from the point *E* let there be drawn the straight line *EZ* parallel to the straight line *AD*. And since the straight line *ZE* is parallel to the straight line *AD* and on them *BED* falls, the angle *ZED* is equal to the angle *ADG*. Therefore [the angle] *ZED* is also given. Since from the straight line *BG* given in position and from the given point *E* there is drawn the straight line *EZ* making a given angle *ZEG*, *EZ* is given in position. Since from a given point [*A*] there is drawn the straight line *AD* parallel to the given straight line *ZE*, *AD* is given in position.

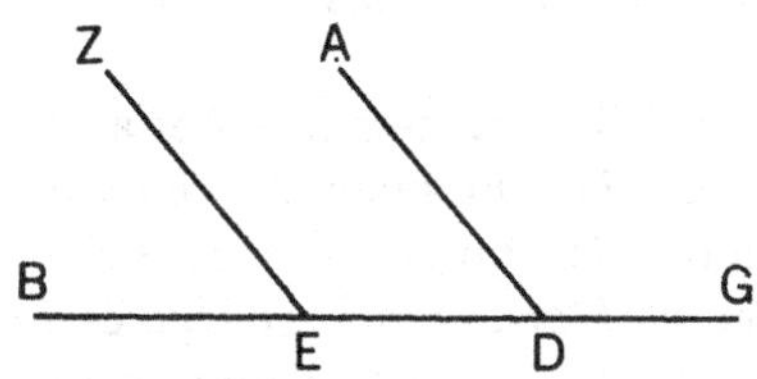

Another proof

On *BG* let there be taken an arbitrary point *E* and let *E* and *A* be joined. Since the point *A* is given, *AE* is given in position. *BG* is also given in position. Therefore the angle *AED* is given. And the angle *ADE* is also given. Therefore the remaining angle *EAD* is also given. Since from the straight line *EA* given in position and from the point *A* in it there is drawn the straight line *AD* making a given angle *EAD*, *AD* is therefore given in position.

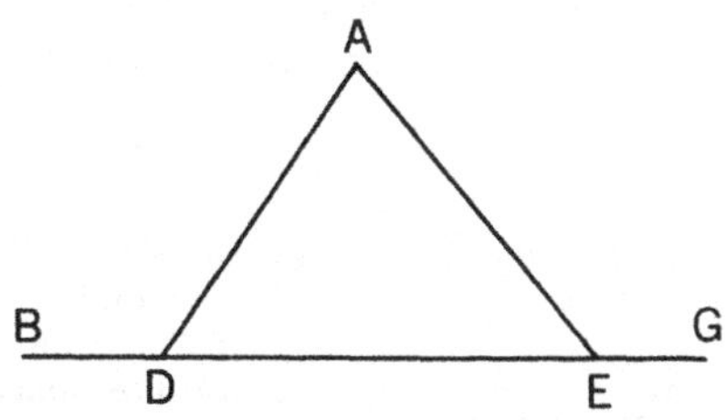

XXXI. If from a given point to a straight line given in position there is drawn a straight line given in magnitude, that line is also given in position.

From a given point *A* to the straight line *BG* given in position let there be drawn the straight line *DA* given in magnitude. I say that it [*DA*] is also given in position.

Centro enim *A* spatioque *AD* circulus describatur *EDZ*. Positione ergo *EDZ* circulus. Datum enim est *A* centrum positione, et e centro recta *AD* magnitudine. Positione vero et *BG* recta. Si autem due linee positione date se invicem secent, datus est punctus ad quem se intersecant. Datus ergo *D* punctus. Estque et *A* punctus datus. Positione ergo est et *AD*.

XXXII. Si in equidistantibus positione datis rectis recta linea protrahatur datos faciens angulos, data est ea protracta magnitudine.

In equidistantibus enim positione datis rectis *AB*, *GD* recta linea protrahatur *EZ* datos faciens angulos *BEZ* et *EZD*. Dico quod data est *EZ* magnitudine.

Sumatur enim super *GD* datus punctus *I* et a puncto *I* recte *EZ* equidistans protrahatur recta *IT*. Quoniam ergo equidistans est *IT* recta recte *EZ* eisque incidit recta *GD*, equalis est ergo *EZD* angulus angulo *TID*. Datus est autem angulus *EZD*. Datus est ergo et *TID* angulus. Quoniam ergo a positione data recta et in ea dato puncto *I* recta linea tracta est *IT* datum faciens angulum *TIZ*, positione ergo est recta *IT*. Positione autem et *AB*. Datus est ergo punctus *T*. Est autem et *I* punctus datus. Data est ergo *IT* recta magnitudine. Est autem ei equalis *EZ* recta. Data est ergo et *EZ* recta magnitudine.

2 enim *tr. P post* est / recta *om. B*
2–3 Da[tum....recta *illeg. D*
3 Si autem *bis D*
4–5 se[1]....punctus[1] *illeg. D*
5 *post* D *add. mg. O* ex XXV / est et AD *illeg. D* / *post* AD *add. mg. O* ex XXVI
6 XXXII: 32[a] *B*
6–7 recta .... fa]ciens *illeg. D*
7–8 magnitudine .... linea *illeg. D*
9–11 Di[co....Sumatur *illeg. D*
11–12 equi[distans....equi]distans *illeg. D*
12 IT[2]: ut *D*
13–14 est....Datus[1] *illeg. D*
14 est autem: autem est *B* / EZD: EZ *B* / *de* Datus[2] *add. mg. D* ex propositione / est[2] *om. B*
15 recta....re]cta *illeg. D*
16 *post* angulum *add. mg. O* ex XXIX / ergo *tr. DO post* est / *de* ergo *add. mg. D* ex 29
16–17 IT....est[1] *illeg. D*
17 ergo *om. B* / *supra* T *add. O* ex XXV
18 magnitudine....equalis *illeg. D*
19 et *om. DO* / *post* recta *add. mg. O* ex XXVI

With the center *A* and the [radial] distance *AD* let there be described the circle *EDZ*. Then the circle *EDZ* is given in position. For the center *A* is given in position. And the straight line *AD* from the center is given in magnitude. But the line *BG* is also given in position. If two lines given in position intersect each other, the point at which they intersect is given. Therefore the point *D* is given. And the point *A* is also given. Therefore *AD* is also given in position.

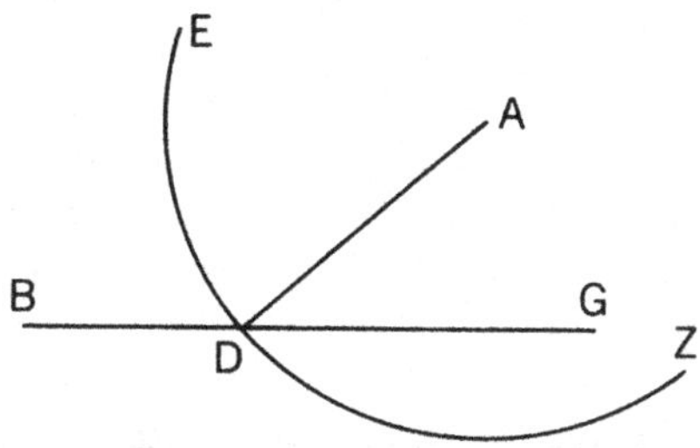

XXXII. If between parallel straight lines given in position there is drawn a straight line making given angles, the line so drawn is given in magnitude.

Between the parallel straight lines *AB* and *GD* given in position, let there be drawn the straight line *EZ* making the given angles *BEZ* and *EZD*. I say that *EZ* is given in magnitude.

On *GD* let there be taken a given point *I* and from the point *I* let there be drawn the straight line *IT* parallel to *EZ*. Since *IT* is parallel to *EZ* and the straight line *GD* intersects them, the angle *EZD* is equal to the angle *TID*. But the angle *EZD* is given. Therefore the angle *TID* is also given. Since from the straight line given in position and from the given point *I* in it there is drawn the straight line *IT* making the given angle *TIZ*, the straight line *IT* is given in position. But *AB* is also given in position. Therefore the point *T* is given. But the point *I* is also given. Therefore the straight line *IT* is given in magnitude. But the straight line *EZ* is equal to it [*IT*]. Therefore the straight line *EZ* is also given in magnitude.

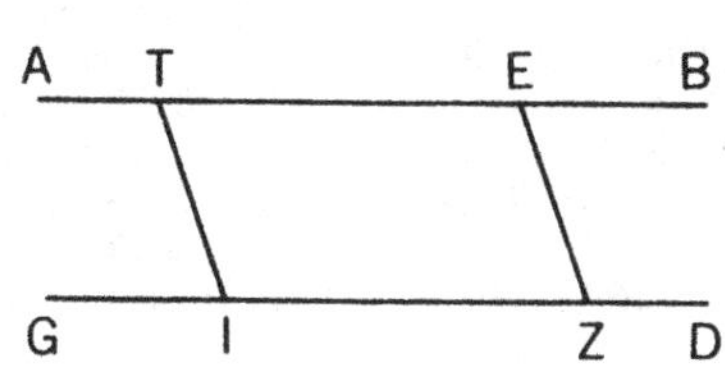

XXXIII. Si in equidistantibus positione datis rectis recta linea protrahatur data magnitudine, datos faciet ea angulos.

In equidistantibus enim positione datis *AB*, *GD* recta linea trahatur *EZ* data magnitudine. Dico quod datos faciet ea angulos *BEZ* et *EZD*. Recipiatur enim super *AB* datus punctus *I* et ab eo recte *EZ* equidistans protrahatur recta *IT*. Equalis est ergo recta *ZE* recte *IT*. Data est autem (62v) *EZ* magnitudine./Data ergo et *IT*. Et *I* punctus datus est. Centro ergo *I* spatioque *IT* circulus descriptus erit positione. Describatur sitque *KTL*. Positione ergo est *KTL*. Positione autem et *GD*. Datus est ergo *T* punctus. Estque et *I* punctus datus. Positione ergo est recta *IT*. Positione autem et *GD* linea. Datus ergo est et *ITD* angulus. Estque etiam angulo *EZD* equalis. Datus ergo est et angulus *EZD*. Et reliquus ergo *ZEB* datus.

Aliter

Sumatur super *GD* datus punctus *I* iaceatque recte *EZ* equalis *ID*. Et centro *I* quidem spatio vero *ID* circulus describatur *DB*. Positione ergo est *DB* circulus. Datum autem est centrum eius positione lineaque e centro magnitudine. Positione autem et *AB*. Datus ergo est *B* punctus. Est autem et *I* datus. Positione ergo *BI*. Positione autem et *GD*. Datus ergo est et *BID* angulus. Et si equidistans est *EZ* recta recte *BI*, erit etiam *EZI* angulus

1 XXXIII: 33ᵃ *B*
1–2 linea protrahatur *illeg. D*
3 equidistantibus....positione *et* da[tis....traha]tur *illeg. D*
4 faciet *tr. B post* ea / BEZ:BZ *D* / EZD:IEZ *B*
5 Recipiatur....datus *illeg. D*
6 *de* est[1] *add. mg. D* ex propositione / recta[2]....Data *illeg. D* / ZE:EZ *B*
7 *ante* ergo[1] *scr. P* est / *post* I[1] *scr. B* et L
7–8 Centro....spatio]que *illeg. D*
8 *supra* erit *add. DO* ex diffinitione
9 est....Datus *illeg. D* / est[1] *om. P* / *post* est[1] *scr. B* et / *supra* T *add. O* ex XXV
10 *de* ergo *add. mg. D* ex 26 / *supra* IT *add. O* ex XXVI
10–11 IT....est *illeg. D*
11 ITD *tr. DO post* angulus / ITD: IT *B*
12 angulus *tr. B post* EZD
12–13 reliquus....Aliter *illeg. D*
15 I....cir[culus *illeg. D* / quidem: quod *B* / *de* Positione *add. mg. D* ex diffinitione
15–16 Positione....DB *illeg. D*
16 autem *tr. D post* est / eius.... centro *illeg. D* / e centro: EZ *B*
17 ergo est: est igitur *B* / *post* autem[2] *scr. B* A
17–18 Datus....GD *om. P* / autem[2] ....Positione[2] *illeg. D*
18 *de* Positione[1] *add. mg. O* ex XXVI / *de* Datus *add. mg. O* ex XXV
19 recta....an]gulus *illeg. D*

XXXIII. If between parallel straight lines given in position there is drawn a straight line given in magnitude, that line will make given angles with the parallel lines.

Between parallel straight lines *AB* and *GD* given in position, let there be drawn a straight line *EZ* given in magnitude. I say that it [*EZ*] will make the given angles *BEZ* and *EZD*.

On *AB* let there be taken a given point *I* and from it let there be drawn the straight line *IT* parallel to *EZ*. Then $ZE = IT$. Now *EZ* is given in magnitude. Therefore *IT* is also given [in magnitude]. And the point *I* is given. Therefore the circle described with the center *I* and the radius *IT* will be given in position. Let is be drawn and be *KTL*. Therefore *KTL* is given in position. Now *GD* is also given in position. Therefore the point *T* is given. But the point *I* is also given. Therefore the straight line *IT* is given in position. Now the line *GD* is also given in position. Therefore the angle *ITD* is also given. And it is also equal to the angle *EZD*. Then the angle *EZD* is also given. Therefore the remaining angle *ZEB* is also given.

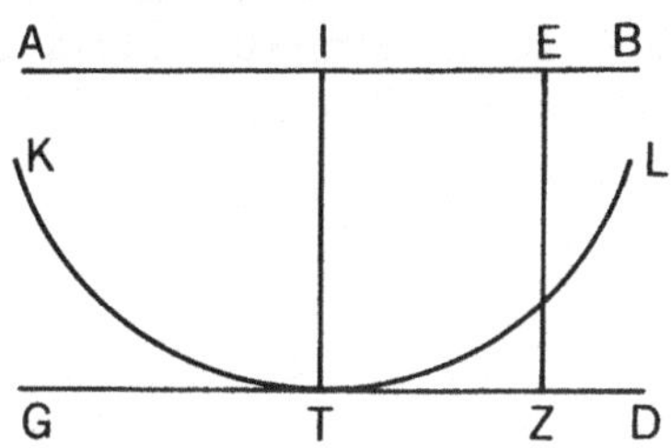

Another proof

Let there be taken on the straight line *GD* a given point *I*, and let *ID* be put equal to *EZ*. And with the center *I* and the radius *ID*, let there be

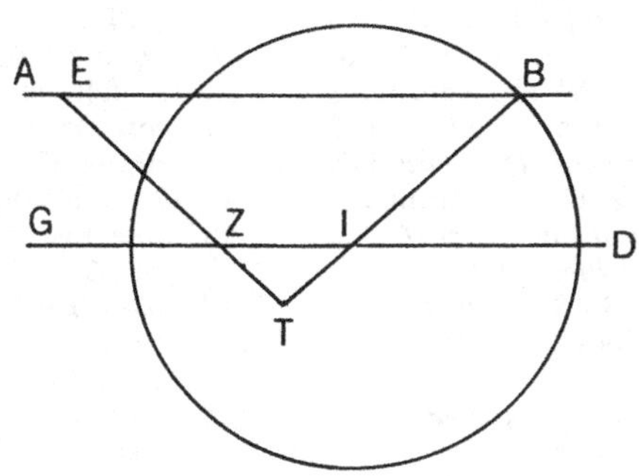

datus. Quare et reliquus *ZEB* angulus datus est. Si autem non, conveniant *EZ*, *IB* ad punctum *T*. Quoniam ergo equalis est *EZ* recta recte *DI*, hoc est, *IB* et est equidistans recta *EB* recte *ZI*, equalis ergo est et *ZT* recta recte *TI*. Quare et angulus *TIZ* angulo *TZI* est equalis. Datus autem et angulus *TIZ*. Datus ergo et angulus *IZT*. Quare et qui deinceps *IZE* angulus datus est. Et reliquus *ZEB* datus est.

XXXIIII. Si in equidistantibus positione datis rectis a dato puncto recta linea protrahatur, in datam proportionem dividetur.

In equidistantibus enim positione datis rectis *AB*, *GD* a dato puncto *E* recta linea protrahatur *EZI*. Dico quod proportio est recte *EZ* ad rectam *ZI* data.

Protrahatur enim ab *E* puncto super rectam *GD* cathetus *EKT*. Quoniam ergo a dato puncto *E* super positione datam rectam *GD* recta linea protracta est datum faciens angulum *ETI*, positione ergo est *ET*. Positione autem et utraque rectarum *AB*, *GD*. Datum ergo est et utrumque punctorum *K* et *T*. Est autem et *E* punctus datus. Data ergo utraque *EK*, *KT*. Proportio ergo recte *EK* ad *KT* rectam data. Estque sicut *EK* ad *KT* ita *EZ* ad *ZI*. Proportio ergo et recte *EZ* ad rectam *ZI* data.

1–2 conveni[ant. . . .pun]ctum *illeg. D*
2 EZ, IB: ZIB *B* / est *om. B*
3 I[B. . . .equa]lis *illeg. D* / *post* ergo *add. mg. O* ex II V[ti] elementorum / est[3] *tr. B ante* ergo / et ZT: IZT *B*
4 TIZ. . . .Datus *illeg. D* / *supra* TIZ angulo *add. mg. O* ex V primi elementorum / *post* autem *scr. OP* et
5–6 IZ[E. . . .est[2] *illeg. D*
7 XXXIIII: 34[a] *B* / Si: Cum *D*
7–8 equidistan[tibus . . . . li]nea *illeg. D*
8–10 dat[am . . . . recta *illeg. D*
10 *post* quod *scr. B* et
10–12 quod. . . .puncto *illeg. D*
12 cathetus: catetus *B hic et ubique* / EKT . . . . linea *illeg. D*
14 *supra* est ET *add. O* ex XXX
14–15 fa[ciens. . . .Da]tum *illeg. D*
15 *post* Datum *add. mg. O* ex XXV / et[2] *om. D*
15–17 pun[ctorum. . . .re]cte *illeg. D*
16 *ante* EK *scr. B* et / *post* KT *add. mg. O* ex XXVI
17 *supra* EK ad[1] *add. O* per I / sicut *bis P*
17–18 rec[tam. . . .ad[2] *illeg. D*
18 *supra* EZ ad *add.* O ex II, VI elementorum

described the circle DB. Then the circle DB is given in position. Now its center [I] is given in position and the line from the center [radius BI] is given in magnitude. But AB is also given in position. Therefore the point B is given. Now I is also given. Therefore BI is given in position. Now GD is also given in position. Therefore the angle BID is also given. And if EZ is parallel to BI, the angle EZI will also be given. Hence the remaining angle ZEB is also given. But if [EZ is] not [parallel to BI], EZ and IB would meet at point T. Since the straight line EZ is equal to the straight line DI, that is, to IB, and the straight line EB is parallel to the straight line ZI, therefore $ZT = TI$. Hence also $\angle TIZ = \angle TZI$. Now the angle TIZ is also given. Therefore the angle IZT is also given. Hence the adjacent angle IZE is also given. And the remaining angle ZEB is given.

XXXIV. If to parallel straight lines given in position there is drawn a straight line from a given point, that straight line will be divided in a given ratio.

To the parallel straight lines AB and GD let there be drawn a straight line EZI from the given point E. I say that the ratio $EZ:ZI$ is given.

From the point E to the straight line GD, let there be drawn a perpendicular line EKT. Since from the given point E to the straight line GD given in position there is drawn a straight line [ET] making the given angle ETI, hence ET is given in position. But each of the straight lines AB and GD is also given in position. Therefore each of the points [of intersection] K and T is also given. But the point E is also given. Therefore both EK and KT are given. Therefore the ratio $EK:KT$ is given. And $EK:KT = EZ:ZI$. Therefore the ratio $EZ:ZI$ is given.

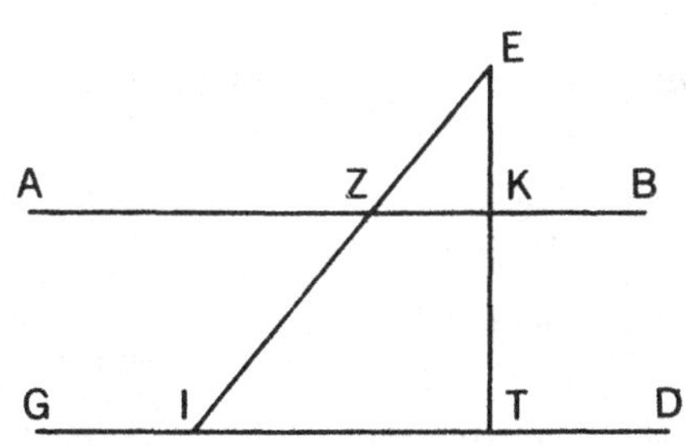

Aliter

In equidistantibus positione datis rectis *AB*, *GD* a dato puncto *E* recta linea trahatur *EZI*. Dico quod proportio est recte *IE* ad *EZ* data.

Protrahatur enim ab *E* puncto super *GD* cathetus *ET*. Et eiciatur super *K*. Quoniam ergo a dato puncto *E* super positione datam rectam *GD* recta linea tracta est *ET* datum faciens angulum *ETI*, positione ergo est *TEK*. Positione autem et utraque *AB*, *GD*. Datus ergo est uterque punctus *T*, *K*. Est autem et *E* datus. Data ergo utraque *TE* et *EK*. Proportio ergo recte *TE* ad *EK* data. Sicut autem *TE* ad *EK* ita *IE* ad *EZ*. Proportio ergo recte *IE* ad *EZ* data.

XXXV. Si a dato puncto super positione datam rectam recta linea protrahatur dividaturque in datam proportionem et per punctum sectionis secundum positione datam rectam recta linea trahatur, data est ea tracta positione.

A dato enim puncto *A* super positione datam rectam *GD* recta linea trahatur *AD* dividaturque in datam proportionem linee *DE* ad *EA* et trahatur per punctum *E* recte *BG* equidistans *ZEI*. Dico quod positione est recta *ZEI*.

Trahatur a puncto *A* super *BG* cathetus *AT*. Quoniam ergo a dato puncto *A* super positione datam rectam *BG* recta linea tracta est *AT* datum faciens angulum *ATD*, positione ergo est recta. At positione autem et *BG*. Datus ergo *T* punctus. Est autem et *A* datus. Data ergo est *AT*. Et quoniam pro-

1–2 Aliter....recta *illeg. D*
3–4 traha[tur....GD *illeg. D*
4 ET:T *B*
4–6 Et....linea *illeg. D*
6 est^{2} *om. B* / *supra* TEK *add. O* ex XXX
6–7 datum....uterque *illeg. D*
7 ergo *tr. B post* est / *post* uterque *add. mg. O* per XXV
8 *post* ergo^{1} *scr. B* et / *supra* et EK *add. O* per XXVI
8–9 E[st....Sicut *illeg. D*
9 *post* data *add. mg. O* per primam / *supra* ita *add. O* per IIII, VI elementorum
9–10 ad^{2}....data *illeg. D*
11 XXXV: 5^{a} *B*
11–12 a . . . dividaturque *illeg. D*
12–13 da[tam . . . . datam *illeg. D*
13 linea *om. B*
13–16 recta . . . . linee *illeg. D*
15 datam *tr. B post* rectam
16 trahatur: tracta *B*
17–19 BG....Trahatur *illeg. D*
19 *post* Trahatur *scr. B* autem
19–20 Quoniam....rectam *illeg. D*
20 linea....est *illeg. D*
21 angulum *om. B* / est *om. B* / *supra* recta *add. O* per XXX
21–22 angulum....T *illeg. D*
22 *supra* T punctus *add. O* per XXV / *supra* Data *add. O* ex XXVI
22–1(p.s.) ergo^{2}....sicut *illeg. D*

Another proof

Within the parallel straight lines *AB* and *GD* given in position let there be drawn from the point *E* the straight line *EZI*.* I say that the ratio *IE*:*EZ* is given.

From the point *E* to *GD* let there be drawn the perpendicular line *ET* and let it be projected to point *K*. Since from the given point *E* to the straight line *GD* given in position there has been drawn the straight line *ET* making a given angle *ETI*, [the straight line] *TEK* is given in position. But both *AB* and *GD* are also given in position. Therefore each point of *T* and *K* is given. Now *E* is also given. Therefore each of the straight lines *TE* and *EK* are given. Therefore the ratio *TE*:*EK* is given. But *TE*:*EK* = *IE*:*EZ*. Therefore the ratio *IE*:*EZ* is given.

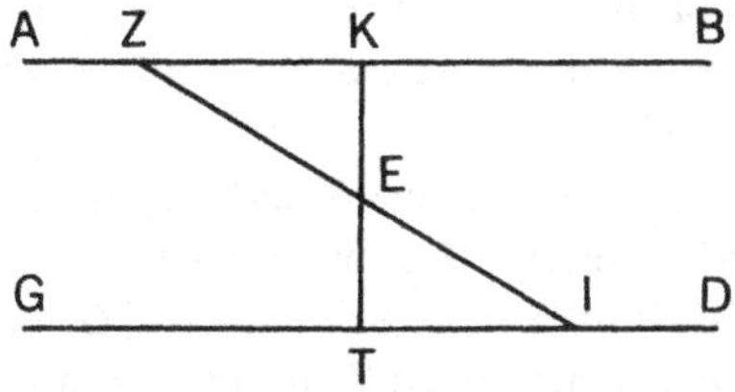

XXXV. If from a given point to a straight line given in position there is drawn a straight line, and if it is divided according to a given ratio, and if through the point of intersection there is drawn a straight line parallel to the straight line given in position, that line drawn [parallel] is given in position.

From the given point *A* to the straight line *GD* given in position, let there be drawn the straight line *AD* and let it be divided in the given ratio of *DE* to *EA* and let there be drawn through point *E* a straight line *ZEI* parallel to *BG*. I say that *ZEI* is given in position.

From the point *A* to *BG*, let there be drawn the perpendicular *AT*. Since from the given point *A* to the straight line *BG* given in position there has been drawn a straight line *AT* making the given angle *ATD*, that straight line is given in position. But *BG* is also given in position. Therefore the

* *i.e.* through the point *E* the straight line *ZEI*.

portio est recte *DE* ad *EA* data <et> sicut autem *DE* ad *EA* ita *TK* ad *KA*, proportio ergo et recte *TK* ad *KA* data. Et componenti proportio recte *TA* ad *AK* data. Data autem *TA*. Data ergo et *AK*, sed et positione. Et *A* pnuctus datus est. Datus ergo et *K*. Quoniam ergo per datum punctum *K* secundum positione datam rectam *BG* recta linea tracta est *ZI*, positione ergo est *ZI*.

XXXVI. Si a dato puncto super positione datam rectam recta linea trahatur et apponatur ei recta proportionem habens ad ipsam datam, per terminum autem adiuncte secundum positione datam recta linea trahatur, data est illa protracta positione.

(63r) /A dato enim puncto *A* super positione datam rectam *BG* recta linea trahatur *AD* et coniungatur recte *AD* recta *AE* proportionem habens ad *AD* datam. Per punctum autem *E* recte *BG* equidistans trahatur *ZK*. Dico quod positione est *ZK*.

Trahatur enim a puncto *A* super *BG* cathetus *AT* et protrahatur super *I*. Quoniam ergo a dato puncto *A* super positione datam rectam *BG* recta linea tracta est *AT* datum faciens angulum *ATG*, positione ergo est *TAI*. Positione autem et *BG*. Datus ergo *T* punctus. Est autem et *A* datus. Data ergo est recta *AT*. Et quoniam proportio recte *DA* ad *AE* data, sicut autem *DA* ad *AE* ita *TA* ad *AI*, proportio ergo et recte *TA* ad *AI* data. Data au-

1 *post* EA^2 *bis B* data sicut autem DE ad EA / *de* TK *add. mg. D* per 2, 6^{ti} elementorum / KA:AK *B*
2 propor[tio....propor]tio *illeg. D* / componenti *corr. B ex* componente / *supra* recte^2 *add. O* per VI
3 *supra* Data^2 *add. O* per II
3–4 Data^2....datus *illeg. D*
4 *post* Datus *add. mg. O* per XXVII
5 K....recta *illeg. D*
6 *de* ergo *add. mg. D* per 28 / *supra* ZI *add. O* per XXVIII
7 XXXVI *om. B* / Si . . . . puncto *illeg. D*
8 et . . . . recta *illeg. D* / habens *tr. B post* ipsam
9 autem . . . positio]ne *illeg. D* / recta: rectam *D*
10 pro[tracta positione *illeg. D*
12 AD et *illeg. D*
13 datam *tr.* B *ante* ad *ln. 12*/ autem: a *P* / ZK:CK *P*/pun[ctum....equi]distans *illeg. D*
14 est *om. P*
15 A....et *illeg. D* / super^1 *om. B*
16–17 rectam....tracta *illeg. D*
17 *post* est^2 *scr. D* recta et(?)/ *supra* TAI *add. O* ex XXX
18 Positione....Datus *illeg. D* / et^1 *om. B* / *post* T *add. mg. O* per XXV
19 ergo *tr. BOP post* est / recta *om. D* / *supra* recta *add. O* per XXVI / DA....data *illeg. D* / autem *om. P*
20 DA *corr. D ex* DE / AE *corr. D ex* EA / TA *corr. D ex* TK / AI^1 *D corr. ex* KA / *ante* proportio *scr. D* data et componenti / ergo....AI *illeg. D*

point *T* is given. But *A* is also given. Therefore *AT* is given. Since the ratio *DE*:*EZ* is given and *DE*:*EA* = *TK*:*KA*, the ratio *TK*:*KA* is also given. And, by composing, the ratio *TA*:*AK* is given. Now *TA* is given. Therefore *AK* is also given [in magnitude], but [it is given] also in position. And the point *A* is given. Therefore *K* is also given. Since through the given point *K* there is drawn the straight line *ZI* [parallel] to *BG* given in position, hence *ZI* is given in position.

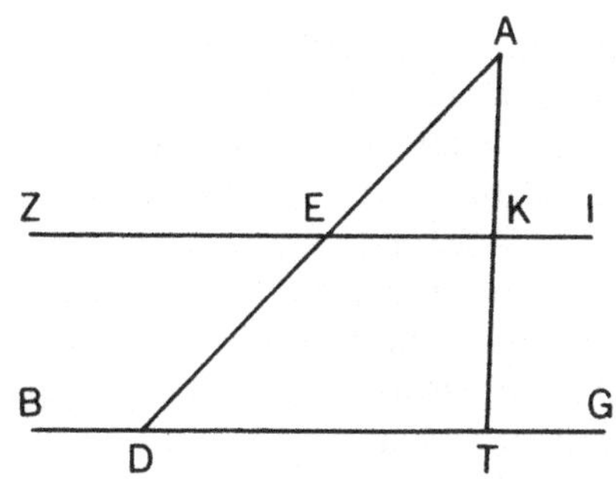

XXXVI. If from a given point there is drawn a straight line to a straight line given in position and to it [*i.e.* the drawn line] a straight line is added having a given ratio to the same, and if through the extremity [of the added line] there is drawn a straight line [parallel] to the line given in position, that line drawn [parallel] is given in position.

From the given point *A* to the straight line *BG* given in position, let there be drawn a straight line *AD* and let there be joined to *AD* the straight line *AE* having a given ratio to *AD*. Through the point *E*, let there be drawn a straight line *ZK* parallel to *BG*. I say that *ZK* is given in position.

Let there be drawn through the point *A* to *BG* the perpendicular *AT* and let it be protracted to point *I*. Since from the given point *A* to the straight line *BG* given in position there is drawn a straight line *AT* making the given angle *ATG*, therefore *TAI* is given in position. But *BG* is also given. Therefore the point *T* is given. Now *A* is also given. Therefore the straight line *AT* is given. Since the ratio *DA*:*AE* is given and *DA*:*AE* = *TA*:*AI*, the ratio *TA*:*AI* is also given. Now the straight line *TA* is given. Therefore *AI* is also given [in magnitude], but [it is given] also in position. The point *A* is given. Therefore *I* is also given. Since through the given point *I* there is

tem recta *TA*. Data ergo et *AI*. Sed et positione. Et datus est punctus *A*. Datus ergo et *I*. Quoniam ergo per datum punctum *I* secundum positione datam rectam *BG* recta linea tracta est *ZIK*, positione ergo est *ZIK*.

XXXVII. Si in equidistantibus positione datis rectis recta linea trahatur dividaturque in datam proportionem, per punctum autem sectionis secundum positione datas rectas recta linea protrahatur, data est ea protracta positione.

In equidistantibus enim positione datis rectis *AB*, *GD* recta linea trahatur *EZ* dividaturque in datam proportionem recte *ZI* ad *IE* et protrahatur per *I* ambarum *AB*, *GD* equidistans *TK*. Dico quod positione est *TK*.

Sumatur enim super *AB* datus punctus *L*. Trahaturque deorsum a puncto *L* super *GD* cathetus *LN*. Et quoniam a dato puncto *L* super positione datam rectam *GD* recta linea tracta est *LN* datum faciens angulum *IND*, positione ergo est *LN*. Positione autem et *GD*. Datus ergo *N* punctus. Est autem et *L* punctus datus. Data ergo est *LN*. Et quoniam proportio est recte *ZI* ad *IE* data, sicut autem *ZI* ad *IE* ita *NM* ad *ML*, proportio ergo et recte *MN* ad *ML* data. Quare et recte *LN* ad *ML* proportio est data. Data autem et *NL*. Data ergo et *LM*, sed et positione. Et datus est punctus *L*. Datus ergo et *M*. Quoniam ergo per datum punctum *M* secundum positione datam rectam *GD* recta linea tracta est *TK*, positione ergo est recta *TK*.

1 recta *om. D* / AI *corr. D ex* AK / *supra* AI *add. DO* per II / et^{2} *om. P* / Et *om. B* / datus est *om. D* / A *tr. D ante* punctus
2 ergo^{1}....per *illeg. D* / et *om. B* / *ante* I^{2} *scr. et del. D* K
3 datam rectam: data recta *B* / ZIK^{1}: ZK *B* / *post* ZIK^{2} *add. mg. O* per XXVIII
3–4 est^{1} .... equidist]antibus *illeg. D*
5 divi[daturque .... data]m *illeg. D*
6 da[tas .... protra]hatur *illeg. D*
8 equidistantibus....recta *illeg. D* / linea *tr. B post* trahatur
9–10 propor[tionem....GD *illeg. D*
11 Sumatur....deor]sum *illeg. D*
12–13 Et....recta *illeg. D*
13 est *om. B*
13–14 faciens....ergo^{2} *illeg. D*
15 datus *om. B* / *supra* LN *add. O* per XXVI
15–16 pun[ctus....data *illeg. D*
16 *supra* NM ad ML *add. O* per II VI^{ti} elementorum ducta
16–17 proportio....data^{1} *om. P*
17 *supra* recte^{1} *add. O* EN exagono / *supra* recte^{2} *add. O* per VI / LN: NL *P*
17–18 data^{2}....punct]us *illeg. D*
18 Et *om. B*
19 *de* Datus *add. mg. D* per 27 / *supra* M^{1} *add. O* per XXVII
19–20 per....tr]acta *illeg. D*
20 TK^{1}:KT *B* / est recta *om.* B / *post* TK^{2} *scr. B* reliqua

drawn the straight line *ZIK* [parallel] to the straight line *BG* given in position, therefore *ZIK* is given in position.

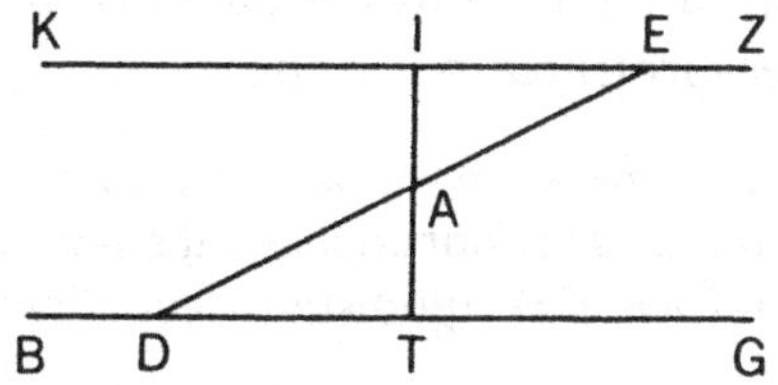

XXXVII. If between parallel straight lines given in position there is drawn a straitht line and it is divided in a given ratio, and if through the point of intersection there is drawn a straight line parallel to the straight lines given in position, that [parallel] line so drawn is given in position.

Between the parallel straight lines *AB* and *GD* given in position, let there be drawn a straight line *EZ* and let it be divided into the given ratio *ZI*:*IE*, and through *I* there is protracted *TK* parallel to both *AB* and *GD*. I say that *TK* is given in position.

Let a given point *L* be taken on the line *AB* and let a perpendicular *LN* be drawn down from the point *L* to *GD*. Since from the given point *L* to the straight line *GD* given in position there is drawn the straight line *LN* making the given angle *LND*, therefore *LN* is given in position. Now *GD* is also given in position. Therefore the point *N* is given. But *L* is also given. Therefore *LN* is given. Since the ratio *ZI*:*IE* is given and $ZI:IE = NM:ML$, therefore the ratio *MN*:*ML* is given. Hence the ratio *LN*:*ML* is given. But *NL* is given. Therefore *LM* is given [in magnitude], but [it is] also in position. And the point *L* is given. Therefore *M* is also given. Since through the given point *M* there is drawn the straight line *TK* parallel to the straight line *GD* given in position, therefore *TK* is given in position.

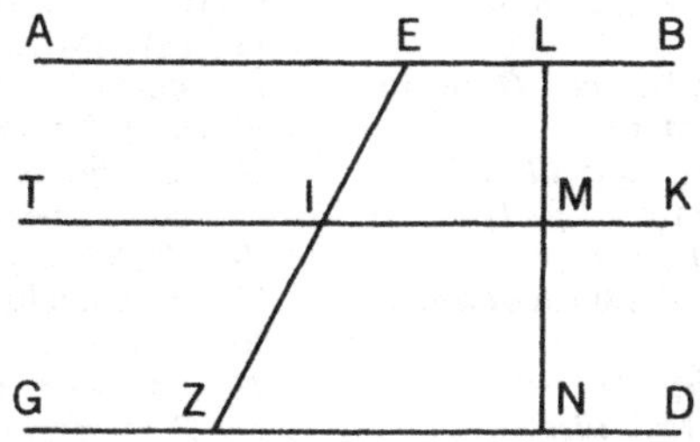

XXXVIII. Si in equidistantibus positione datis recta linea trahatur apponaturque ei recta proportionem habens ad ipsam datam, per terminum autem secundum positione datas equidistans recta linea trahatur, data est ea protracta positione.

In equidistantibus enim positione datis rectis *AB*, *GD* recta linea trahatur *EZ* apponaturque ei recta *EI* proportionem habens ad *EZ* datam. Per punctum autem *I* ambarum *AB*, *GD* equidistans recta linea trahatur *TK*. Dico quod positione est *TK*.

Sumatur enim super *AB* datus punctus *N* protrahaturque a puncto *N* super rectam *GD* cathetus recta linea *NM* protrahaturque super *L*. Quoniam ergo a dato puncto *N* super positione datam rectam *GD* recta linea tracta est *NM* datum faciens angulum *NMD*, positine ergo est recta *LNM*. Positione autem et recta *GD*. Datus ergo est punctus *M*. Est autem et *N* datus. Data ergo est recta *NM*. Et quoniam proportio est recte *ZE* ad *EI* data, sicut autem *ZE* ad *EI* ita *MN* ad *NL*, proportio ergo et recte *MN* ad *NL* data. Data autem *NM*. Data ergo et *NL*, sed et positione. Et est *N* datus. Datus ergo et *L*. Quoniam ergo per datum punctum *L* secundum positione datam rectam *AB* recta linea tracta est *TK*, positione ergo est recta *TK*.

XXXIX. Si trianguli unumquodque laterum datum fuerit magnitudine, datus est et triangulus specie.

Trianguli enim *ABG* unumquodque laterum datum esto magnitudine. Dico quod *ABG* trigonus datus est specie.

1 XXXVIII *om. B* / Si . . . . trahatur *illeg. D*
2–3 ei . . . . . secundum *illeg. D*
3–4 equidistans . . . . po]sitione *illeg. D*
4 *ante* positione *add. mg. O* ut in premissa argumentare
5–6 In. . . . app]onaturque *illeg. D*
6–7 EI. . . . equidistans *illeg. D*
7 autem *om. P* / TK:EK *P*
7–9 li[nea. . . . protraha]turque *illeg. D*
9–11 a. . . . ergo *illeg. D*
11–12 puncto. . . . NMD *illeg. D*
12 LNM:LMN *B*
12–14 ergo. . . . rec]ta *illeg. D*
14 ergo *tr. P post* est[1]
14–15 NM. . . . ad[3] *illeg. D*
14–16 Et. . . . NM *om. B*
15 MN[1]:ML *P*
16 data *om. D*
16–18 positione. . . . rectam *illeg. D*
18 rectam: reliqua(?) *corr. B ex* rectam / linea. . . . TK[2] *illeg. D*
19 XXXIX: 39[a] *B*
19–20 unum[quoque . . . . est *illeg. D*
20 et *om. B*
21 esto *corr. B ex* est
21–22 u[numquodque. . . . tri]gonus *illeg. D*

XXXVIII. If within parallel straight lines given in position there is drawn a straight line and to it there is added some other straight line having a given ratio to the first one, and if through the extremity [of the added line] there is drawn a straight line parallel to the parallel lines given in position, that [parallel] line so drawn is given in position.

Within the parallel straight lines AB and GD given in position let there be drawn a straight line EZ and to it let there be added some other straight line EI having a given ratio to EZ. Through the point I let there be drawn the straight line TK parallel to both AB and GD. I say that TK is given in position.

On AB let there be taken a given point N and let there be drawn a perpendicular NM from point N to the straight line GD, and let it be protracted to L. Since from the given point N to the straight line GD given in position there has been drawn the straight line NM making the given angle NMD, hence the straight line LNM is given in position. But the straight line GD is given in position. Therefore the point M is given. But

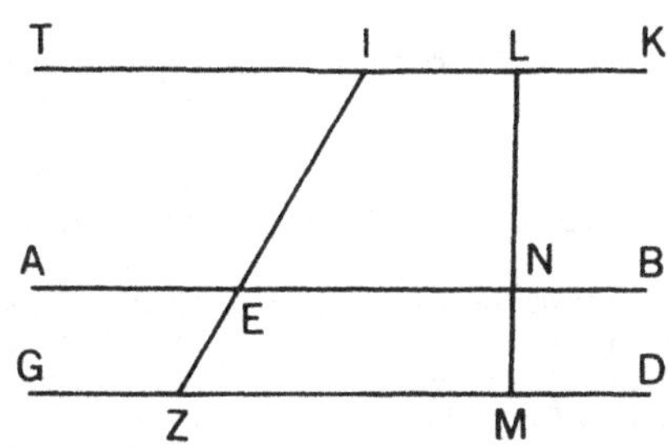

N is also given. Therefore the straight line NM is given. Since the ratio $ZE{:}EI$ is given and $ZE{:}EI = MN{:}NL$, hence the ratio $MN{:}NL$ is given. But NM is given. Therefore NL is given [in magnitude], but [it is given] also in position. And N is given. Therefore L is also given. Since through the given point L there is drawn the straight line TK parallel to the straight line AB given in position, hence the straight line TK is given in position.

XXXIX. If each side of a triangle is given in magnitude, the triangle is also given in species.

Subiaceat enim recta positione *DM* terminata quidem ad punctum *D*, in oppositum vero quantumlibet ducta, ponaturque recte *AB* quidem equalis recta *DE*. Data est autem *AB*. Data est ergo et *DE*, sed et positione. Est autem et punctus *D* datus. Datus ergo et *E*. Recte autem *BG* equalis *EZ*. Data autem *BG*. Data ergo et *EZ*. Sed et positione. Et est datus punctus *E*. Datus ergo et *Z*. Recte autem *AG* equalis esto *ZI*. Data autem *AG*. Data ergo et *ZI*, sed et positione. Et est datus punctus *Z*. Datus ergo et *I*. Et centro quidem *E*, spatio vero *ED* circulus describatur *DKT*. Positione (63v) ergo *DKT*. Rursum centro/quidem *Z* spatioque *ZI* circulus describatur *IKL*. Positione ergo *IKL*. Positione autem et *TKD* circulus. Datus ergo est *K* punctus estque et uterque *E* et *Z* datus. Datum ergo est unumquodque horum *KE*, *EZ*, *ZK* positione et magnitudine. Datus est ergo *KEZ* trigonus specie. Estque equalis et similis *ABG* trigono. Datus est ergo *ABG* trigonus specie.

1 Sub[iaceat. . . .punctum *illeg. D*
2 recte: recta *B* / quidem: que *B*
2–3 ponaturq[ue. . . .est[1] *illeg. D*
3 recta: recte *B* / est[2] *om. BP* / ergo *corr. O ex* autem
4–5 D. . . .autem *illeg. D*
5 et[2]: etiam *P*
5–6 da[tus. . . .Re]cte *illeg. D*
6 ergo *om. B*
7 et[1] *om. B* / et[2]. . . .Datus *illeg. D*
8–9 Positione. . . .centro *illeg. D*
9 *sub* DKT *add. O* per diffinitionem / *post* centro *add. mg. infer. O* Nota quod figura potest dari magnitudine et non dari specie nec positione. Verbi gratia. Sint due linee AB, CD, fiat paralellogramum AD et sint puncta data positione, i.e., immobilia. C et D puncta moveantur semper tamen distent equaliter. Patet quod AD paralellogramum semper fixum in magnitudine eadem sed continue alterat speciem quia anguli illius non sunt in eadem magnitudine manentes. Similiter patet de trigonis in quos dividitur AD. Datur etiam figura specie et non magnitudine ut si intalligamus* quadratum quod continue crescat vel decrescat manens tamen quadratum semper. Potest etiam dari positione et specie nec tamen magnitudine ut si intelligamus inscriptam circulo utpote exagonum equilaterum. Sitque circuli centrum datum positione et crescat vel decrescat continue circulus cum exagono inscripto manente semper equilatero. Tam circulus quam exagonus sint dati specie non tamen magnitudine. Hec sufficiant ad intelligendum etc. (* *post* intelligamus *scr. et del.* inscriptam circulo utpote exagonum)
9–10 DKT. . . .IKL[2] *om. B.*
10 Positione ergo IKL *om. D*
10–11 TKD. . . .K *illeg. D*
11 *supra* K *add. O* per XXV / *de* Datum *add. mg. D* per 26 / est *om. B*
11–12 unum[quodque. . .ma]gnitudine *illeg. D*
12 *post* EZ *add. mg. O* per XXVI / magnitudine: magni angulus A tudine(?) *B* / trigonus *tr. B post* specie *ln. 13*
13–14 si[milis. . . .specie *illeg. D*

Let each side of the triangle *ABG* be given in magnitude. I say that the triangle *ABG* is given in species.

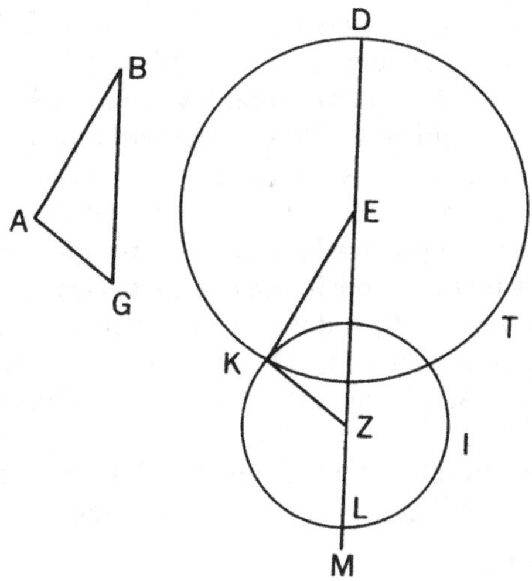

Let a straight line be posited as *DM* given in position, ending in the one point *D* but being drawn in the opposite direction as long as you like, and on it let straight line *DE* be posited as equal to *AB*. But *AB* is given. Therefore *DE* is also given [in magnitude], but [it is given] also in position. And the point *D* is also given. Therefore *E* is also given. Further, [let] *EZ* = *BG*. Now *BG* is given. Therefore *EZ* is also given. But it is given also in position. And the point *E* is given. Therefore *Z* is given. Further, let *ZI* = *AG*. But *AG* is given. Therefore *ZI* is also given, but [it is given] also in position. And the point *Z* is given. Therefore *I* is also given. And, with the center *E* and the radius *ED*, let the circle *DKT* be described. Therefore *DKT* is given in position. Further, with the center *Z* and the radius *ZI*, let the circle *IKL* be described. Therefore *IKL* is given in position. Now the circle *TKD* is given in position. Therefore the point *K* is given and both *E* and *Z* are given. Therefore each of *KE*, *EZ* and *ZK* is given in position and magnitude. Therefore the triangle *KEZ* is given in species. And it is equal and similar to the triangle *ABG*. Therefore the triangle *ABG* is given in species.

XL. Si trianguli unusquisque angulus datus fuerit magnitudine, datus est et triangulus specie.

Trigoni enim *ABG* unusquisque angulus datus esto magnitudine. Dico quod datus est trigonus *ABG* specie.

Ponatur enim positione et magnitudine data recta *DE*, fiantque ad *DE* et ad eius punctum *D* angulo quidem *GBA* equalis angulus rectilineus *EDZ*, angulo autem *BGA* equalis angulus *DEZ*. Reliquus ergo reliquo equalis est qui ad *A* ei qui ad *Z*. Datus autem unusquisque angulorum qui ad *A*, *B*, *G*. Datus ergo et unusquisque angulus qui ad *D*, *Z*, *E*. Quoniam ergo a positione data recta *DE* et in ea puncto dato *D* recta linea tracta est *DZ* datum faciens angulum ad punctum *D*, positione ergo est recta *DZ*. Per eadem vero et recta *EZ* positione est. Datus ergo est *Z* punctus. Est autem et uterque horum *D*, *E* datus. Data ergo unaqueque harum *DZ*, *DE*, *EZ* positione et magnitudine. Datus ergo est *DEZ* trigonus specie. Et est similis trigono *ABG*. Datus est ergo *ABG* trigonus specie.

XLI. Si trigonus unum habet angulum datum, circa vero datum angulum latera adinvicem proportionem habent datam, datus est trigonus ille specie.

1 XL: 40^{a} *B* / magnitudine, datus *illeg. D*
3 enim *om. B*
3–4 esto....quod *illeg. D*
5 positione....re]cta *illeg. D*
6 punctum: puncta *BP* / D *om. BP* / equalis....rectilineus *illeg. D*
7–8 reliquo....ei *illeg. D*
8 autem: et *B*
9 Da[tus....qui *illeg. D*
10–11 recta^{2}....DZ^{1} *illeg. D*
12 eadem: eandem *B* / eadem.... positio]ne *illeg. D* / *post* punctus *add. mg. O* per XXV
13 Data....DZ *illeg. D* / Data: Datus *P* / *supra* harum DZ *add. O* per XXVI
14 DE, EZ: EZ, DE *B* / est *om. B*
14–15 Datus....ABG^{1} *om. D*
16 XLI: 41^{a} *B* / habet *tr. B post* angulum
16–17 da[tum^{2} . . . . latera *illeg. D*
17–1(p.s.) da[tus . . . . unum *illeg. D*

XL. If each angle of a triangle is given in magnitude, the triangle is also given in species.

Let each angle of the triangle *ABG* be given in magnitude. I say that the triangle *ABG* is given in species.

Let straight line *DE* be posited as given in position and magnitude, and to *DE*, at its point *D*, let the rectilinear angle *EDZ* be made equal to the angle *GBA* and the angle *DEZ* equal to the angle *BGA*. Therefore the remaining angle *A* is equal to the remaining one *Z*. Now each of the angles which are at *A*, *B* and *G* is given. Therefore each of the angles which are at *D*, *Z* and *E* is given. Since from the straight line *DE* given in position and from the given point *D* in it* there is drawn the straight line *DZ* making a

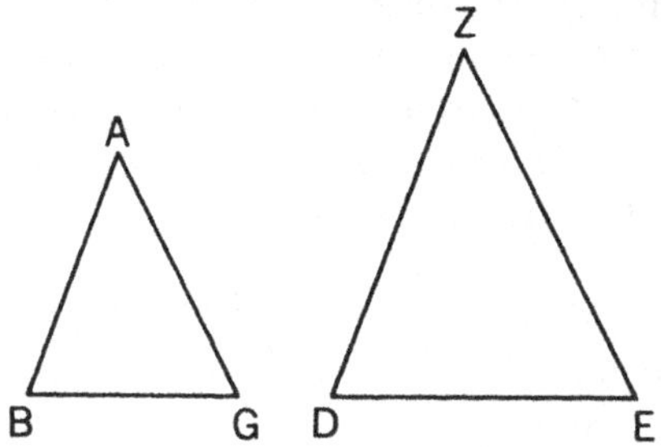

given angle at the point *D*, therefore the straight line *DZ* is given in position. For the same reason, the straight line *EZ* is given in position. Therefore the point *Z* is given. Now both *D* and *E* are given. Therefore each line of *DZ*, *DE* and *EZ* is given in position and magnitude. Therefore the triangle *DEZ* is given in species. And it is similar to the triangle *ABG*. Therefore the triangle *ABG* is given in species.

XLI. If a triangle has one given angle and the sides including the given angle have a given ratio to each other, that triangle is given in species.

* Greek text says "Since on the straight line *DE* given in position and at the given point *D* in it. . . .." In Latin text *a* before *positione* should be *in*, the Greek word of which is *πρός* relating to *δεδομένην ποιοῦσα γωνίαν* (= making a given angle), See Menge, p. 70.

Habeat enim trigonus *ABG* unum angulum datum ad *A*, circa vero angulum qui ad *A* latera *BA*, *AG* proportionem habeant datam. Dico quod *ABG* trigonus datus est specie.

Subiaceat enim positione et magnitudine data recta *DZ* et constituatur ad rectam *DZ* et ad eius punctum *Z* angulo *A* equalis angulus *DZE*. Datus autem angulus *BAG*. Datus ergo et *DZE*. Quoniam ergo a positione data recta *DZ* et in ea dato puncto *Z* recta linea tracta est *ZE* datum faciens angulum *DZE*, positione ergo est recta *ZE*. Et quoniam proportio est recte *BA* ad *AG* data, eadem ei fiat recte *DZ* ad *ZE* et coniungatur *DE*. Proportio ergo et recte *DZ* ad *ZE* data. Data autem *DZ*. Data ergo et *ZE*. Sed et positione. Et est *Z* punctus datus. Datus ergo et *E*. Est autem et uterque horum *D*, *Z* datus. Data ergo unaqueque harum *DZ*, *ZE*, *DE* positione et magnitudine. Datus est ergo *DEZ* trigonus specie. Et quoniam duo trigoni *ABG*, *DEZ* unum angulum uni angulo equalem habent, eum qui ad *A* ei qui ad *Z* et circum equales angulos latera proportionalia, similis ergo est *ABG* trigonus *DEZ* trigono. Datus est autem *DZE* specie. Datus est ergo et *ABG* trigonus specie.

XLII. Si trigoni latera adinvicem proportionem habent datam, datus est trigonus specie.

1 enim *om. B*
2 ad. . . .Dico *illeg. D*
4 Subiace[at. . . .constitua]tur *illeg. D*
5 *ante* rectam *scr. B* datam
5–6 pun[ctum. . . .data *illeg. D* / Datus. . . .DZE *om.*(?) *D*
7 ZE *om. B*
7–8 pun[cto. . . .DZE *illeg. D*
8 *supra* recta *add. O* per XXIX
8–9 proportio. . . .ad^{2} *illeg. D*
9 coniungatur: coniunguntur *B*
9–10 et. . . .data *om. D*
10 *ante* data *bis P* coniungatur DE. Proportio ergo et recte DZ ad ZE / *supra* et ZE *add. O* per II
10–11 Data^{2}. . . .E *illeg. D*
11 *post* E *add. mg. O* per XXVII / autem et *om. D*
12 *supra* harum *add. O* per XXVI
12–13 D[ata. . . .Datus *illeg. D*
13 Datus *om. B*
13–14 trigonus. . . .DEZ *om. D*
14–15 uni. . . .equales *illeg. D*
15 circum: circa *B* cui cum *P*
16 *supra* ergo *add. O* per VI VI^{ti} elementorum
17 est *om. B* / et. . . .specie *illeg. D*
18 XLII: 42^{a} *B*
18–19 trig[oni . . . . tigonus *illeg. D*

Let the triangle *ABG* have one given angle at [the point] *A* and let the sides *BA* and *AG* including the angle at *A* have a given ratio. I say that the triangle *ABG* is given in species.

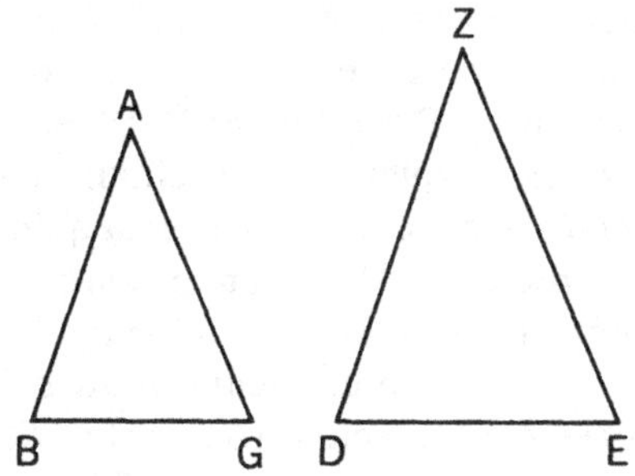

Let a straight line *DZ* be posited as given in position and magnitude, and on the straight line *DZ* at its point *Z* let the angle *DZE* be constructed equal to the angle *A*. Now the angle *BAG* is given. Therefore the angle *DZE* is also given. Since from the given point *Z* in the straight line *DZ* given in position the straight line *ZE* is drawn, forming the given angle *DZE*, therefore the straight line *ZE* is given in position. Since the ratio *BA*:*AG* is given, let the same ratio be made with *DZ* to *ZE* and let *D* and *E* be joined. Then the ratio *DZ*:*ZE* is also given. Now *DZ* is given. Therefore *ZE* is also given. But [it is given] also in position. And the point *Z* is given. Therefore *E* is also given. Now both *D* and *Z* are given. Therefore each of *DZ*, *ZE* and *DE* is given in position and magnitude. Therefore the triangle *DEZ* is given in species. Since the two triangles *ABG* and *DEZ* have one angle equal to one angle, that is, the angle at *A* is equal to the angle at *Z*, and the sides including these equal angles are proportional, therefore the triangle *ABG* is similar to the triangle *DEZ*. But the triangle *DZE* is given in species. Therefore the triangle *ABG* is also given in species.*

XLII. If the sides of a triangle have a given ratio one to the other, the triangle is given in species.

* Note: I have interchanged points *D* and *E* on the figure to make it agree with the text and the figure in the Greek text. This is made necessary by the text which says that the ratio *DZ*:*ZE* is the same as the ratio *BA*:*AG*.

Trigoni enim *ABG* latera adinvicem proportionem habeant datam. Dico quod *ABG* trigonus datus est specie.

Subiaceat enim data magnitudine recta *D*. Et quoniam proportio est recte *AB* ad *BG* data, eadem ei fiat recte *D* ad *E*. Data autem *D*. Data ergo et *E*. Rursum quoniam proportio est recte *BG* ad *GA* data, eadem ei fiat recte *E* ad *Z*. Data autem *E*. Data ergo et *Z*. Et ex tribus rectis que sunt equales tribus datis *D*, *E*, *Z*, quarum due reliqua sunt maiores simul accepte, trigonus fiat *ITK*. Quare equalis erit recta quidem *D* recte *IT*, recta vero *E* recte *TK* et *Z* recte *IK*. Data autem unaqueque harum *D*, *E*, *Z*. Data ergo et unaqueque istarum *IT*, *TK*, *KI* magnitudine. Datus est ergo *ITK* trigonus specie. Et quoniam est sicut *AB* ad *BG* ita *D* ad *E*, equalis autem recta quidem *D* recte *IT*, recta vero *E* recte *TK*, est ergo sicut *AB* ad *BG* ita *IT* ad *TK*. Rursum quoniam est sicut *BG* ad *GA* ita *E* ad *Z*, equalis autem recta quidem *E* recte *TK*, at vero *Z* recte *IK*, est ergo sicut *BG* ad *GA* ita *TK* ad *KI*. Demonstratum autem et sicut *AB* ad *BG* ita *TI* ad *KT*. Per equale ergo sicut *BA* ad *AG* ita *TI* ad *IK*. Similis ergo est *ABG* trigonus *ITK* trigono. Datus est autem *ITK* trigonus specie. Datus ergo et *ABG* specie.

1 *ante* Trigoni *scr. B commentum ad propositionem XXXIX in O* (*cf. p. 118*) Nota quod figura potest dari magnitudine et non dari specie nec positione. Verbi gratia. Sint due linee equidistantes AB, CD, fiat paralellogramum AD et sint AB puncta data positione, id est, immobilia. C et D puncta moveantur semper tamen distant equaliter. Patet quod AD paralellogramum semper fixum in magnitudine eadem sed continue alterat speciem quia anguli illius non sunt in magnitudine eadem manentes. Similiter patet de trigonis in quos dividitur AD. Datur etiam figura specie et non magnitudine ut si intelligamus quadratum quod continue crescat vel decrescat manens tamen quadratum semper. Potest etiam dari positione et specie nec tamen magnitudine ut si intelligamus figuram inscriptam circulo utpote exagonum equilaterum. Sitque circuli centrum datum positione et crescat vel decrescat continue circulus cum exagono inscripto manente semper equilatero. Tam circulus quam exagonus sunt dari specie nec tamen magnitudine. Hec sufficiant ad intelligendum. Insistamus proposito. / *post* enim *scr. B* specie

1–2 Trigo[ni. . . . datus *illeg. D*

2 quod *om. B* / est *om. B*

2–4 s[pecie. . . . da]tą *illeg. D*

3 magnitudine: magnitudo *B*

4–5 fiat. . . . BG *illeg. D*

5–6 eadem. . . . sunt *illeg. D*

7–8 tribus. . . . ITK *illeg. D*

8–9 equ[alis. . . . unaque]que *illeg. D*

9–10 ha[rum. . . . est *illeg. D*

11 *post* specie *add. mg. O* per XXXIX / E:DE *P*

11–12 trigonus. . . . aute]m *illeg. D*

12 E:EI *B*

12–13 quidem. . . . ad² *illeg. D*

13–15 E. . . . ad³ *illeg. D*

14 E *om. B*

15 *post* ita¹ *bis B* E ad Z, equalis autem recta quidem recte TK, at vero Z recte IK, est ergo sicut BG ad G / KI: K ad KI *P*

16–17 BA. . . . tri]gono *illeg. D*

17 est *om. B* / ITK²:K*B* / ABG:DBG *B*

17–1(p.s.) Da[tus¹. . . . u]num *illeg. D*

Let the sides of a triangle ABG have a given ratio to one another. I say that the triangle ABG is given in species.

Let a straight line D be posited as given in magnitude. Since the ratio $AB:BG$ is given, let the same be made of the straight line D to E. But D is given. Therefore E is also given. Further, since the ratio $BG:GA$ is given,

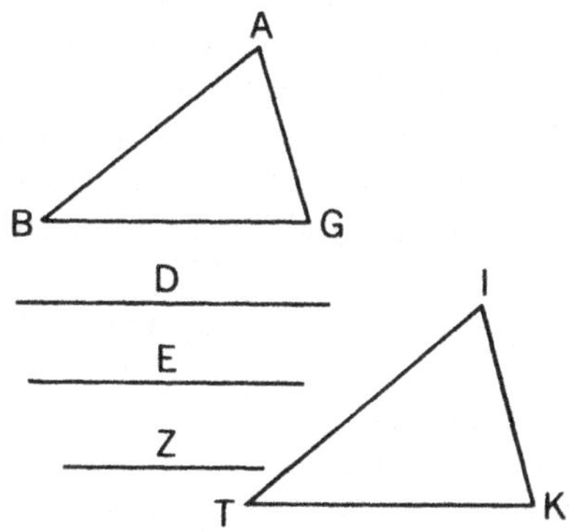

let the same be made of E to Z. But E is given. Therefore Z is also given. And let a triangle ITK be constructed from the three straight lines which are equal to the three given straight lines D, E and Z, any two of which combine to be greater than the remaining line. Hence $D = IT$, $E = TK$ and $Z = IK$. Now each of D, E and Z is given. Therefore each of IT, TK and KI is given in magnitude. Therefore the triangle ITK is given in species. Since $AB:BG = D:E$, $D = IT$ and $E = TK$, therefore $AB:BG = IT:TK$. Further, since $BG:GA = E:Z$, $E = TK$ and $Z = IK$, therefore $BG:GA = TK:KI$. But it has been demonstrated that $AB:BG = TI:KT$. By equality, therefore, $BA:AG = TI:IK$. Therefore the triangle ABG is similar to the triangle ITK. But the triangle ITK is given in species. Therefore the triangle ABG is also given in species.

XLIII. Si trigoni orthogonii circa unum acutum angulum latera proportionem adinvicem habent datam, datus est et trigonus ille specie.

(64r) /Trigoni enim orthogonii *ABG* rectum habentis ad *A* angulum, circa unum acutum angulum *ABG* latera *GB*, *BA* adinvicem habeant proportionem datam. Dico quod datus est *ABG* trigonus specie.

Subiaceat enim positione et magnitudine data recta *DE* et describatur super *DE* semicirculus *DIE*. Positione ergo est *DIE* semicirculus. Et quoniam proportio est recte *GB* ad *BA* data, eadem ei fiat recte *DE* ad *Z*. Proportio ergo et recte *DE* ad *Z* data. Data autem *DE*. Data ergo et *Z*. Et maior est *GB* quam *BA*. Maior ergo *ED* quam *Z*. Adaptetur in circulo recte *Z* equalis *DI* et coniungantur *I*, *E* et centro quidem *D* spatio autem *DI* circulus describatur *TIK*. Positione ergo est *TIK* circulus, datum est enim eius centrum positione et e centro recta magnitudine. Positione autem et *DIE* semicirculus. Datus ergo est *I* punctus. Est autem et uterque istorum *D* et *E* datus. Data ergo est unaqueque istarum *ID*, *DE*, *EI* positione et magnitudine. Datus est ergo *IDE* trigonus specie. Quoniam ergo duo trigoni *ABG*, *DEI* sunt unum angulum uni angulo habentes equalem, eum qui ad *A* ei qui ad *I*, circa vero alios angulos *GBA*, *EDI* latera proportionalia, uterque quoque reliquorum *BGA*, *DEI* simul minor recto, similis ergo est *ABG* triangulus *DEI* triangulo. Datus autem *DIE* trigonus specie. Datus ergo et *ABG* trigonus specie.

1 XLIII *om. B* / orthogonii: ortogonii *B*
2 propor[tionem . . . . datam *illeg. D* / ille *om. B*
4 Trigoni. . . . rect]um *illeg. D* / circa: contra *D*
5 angulum. . . . adinvicem *illeg. D* / GB: BG *B* / habeant *tr. B post* proportionem
7 Subiaceat. . . . magni]tudine *illeg. D* / *post* recta *scr. DO* recte *sed del. O*
8–9 DIE[1]. . . . ad[1] *illeg. D*
9–10 Proportio. . . . data *om. D*
10 *de* Da[ta[2] *add. mg. D* per 2 / ergo[2] *tr. B ante* Data[2]
10–11 Da[ta[2]. . . . GB *illeg. D*
11 ED:DE *B*
12 et[1]. . . . centro *illeg. D*
13 est[2] *om. B*
13–14 datum. . . . positione *illeg. D*
15 Da[tus. . . . uter]que *illeg. D* / est *tr. B ante* ergo
16 *de* Data *add. mg. D* per 26 / ergo *tr. B post* est / *post* unaqueque *add. mg. O* per XXVI
16–17 positione. . . . est *illeg. D*
17 *de* IDE *add. mg. O* per XXXIX
18 angulum. . . . haben]tes *illeg. D*
19–20 la[tera. . . . quo]que *illeg. D*
20 uterque: utrique *B* quoque: quo *B* reliquorum *correxi ex* reliquus *BDOP* / simul: similis *BP* / minor *om. B*
21 *de* similis *add. mg. D* 6, 6[ti] elementorum / *de* ABG *add. mg. O* ex VII VI[ti] elementorum / trian[gulus. . . Datus *illeg. D*/Datus: Datum *B*

XLIII. If the sides including one acute angle of a right triangle have a given ratio to each other, that triangle is also given in species.

Let the sides *GB* and *BA* including one acute angle *ABG* of the right triangle *ABG*, which has a right angle at *A*, have a given ratio to each other. I say that the triangle *ABG* is given in species.

Let the straight line *DE* be posited as given in position and magnitude and let there be described on *DE* a semicircle *DIE*. Therefore the semicircle *DIE* is given in position. Since the ratio *GB*:*BA* is given, let the same ratio be made of *DE* to *Z*. Therefore the ratio *DE*:*Z* is also given. But *DE* is given. Therefore *Z* is also given. And *GB* is greater than *BA*. Therefore *ED* is greater than *Z*. In the circle let *DI* be adapted as equal to *Z* and let *I* and *E* be joined, and with the center *D* and the radius *DI*, let the circle *TIK* be described. Therefore the circle *TIK* is given in position, for its center is given in position and the straight line from the center is given in magnitude. But the semicircle *DIE* is also given in position. Therefore the point *I* is given. Both *D* and *E* are given. Therefore each of *ID*, *DE* and *EI* is given in position and magnitude. Therefore the triangle *IDE* is given in species. Since the two triangles *ABG* and *DEI* have one angle equal to one angle, that is, the angle at *A* is equal to the angle at *I*, and the sides including the angle *GBA* and *EDI* are proportional and each of the remaining angles *GBA* and *DEI* is also less than a right angle at the same time, therefore the triangle *ABG* is similar to the triangle *DEI*. But the triangle *DIE* is given in species. Therefore the triangle *ABG* is also given in species.

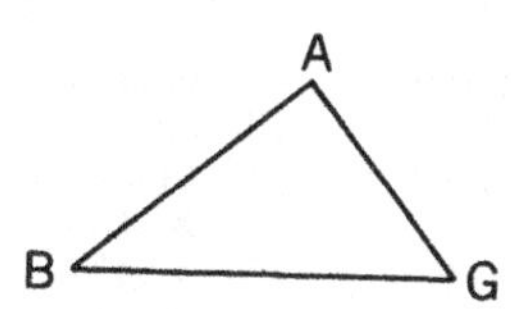

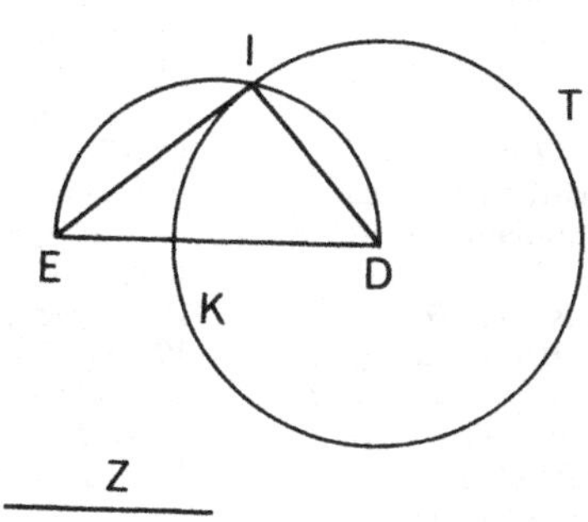

XLIIII. Si trigonus unum angulum habet datum, circa alium vero angulum latera adinvicem proportionem habent datam, datus est trigonus ille specie.

Esto trigonus *ABG* unum habens angulum datum ad *A*, circa alium vero angulum *ABG* latera *AB*, *BG* proportionem habeant datam adinvicem. Dico quod *ABG* trigonus datus est specie.

Non sit enim ad *A* angulus rectus, sed sit primum acutus et trahatur a puncto *B* super *AG* cathetus *BD*. Quoniam datus est *BDA* angulus, est autem et qui ad *A* datus, et reliquus ergo qui ad *B* datus est. Datus est ergo *BAD* trigonus specie. Proportio ergo recte *BA* ad *BD* data. Sed recte *AB* ad *BG* proportio est data. Et recte ergo *BD* ad *BG* proportio est data. Et est rectus angulus *BDG*. Datus est ergo *BDG* trigonus specie. Datus ergo *BGD* angulus. Est autem at *BAG* angulus datus. Et reliquus ergo *ABG* angulus est datus. Datus est ergo *ABG* trigonus specie.

Sed et sit qui ad *A* angulus obtusus producaturque recta *GA* super *E* et trahatur a puncto *B* super *AE* cathetus *BE*. Quoniam ergo datus est angulus *BAG*, et qui deinceps ergo *BAE* datus est. Est autem et *BEA* angulus datus. Et reliquus ergo *EBA* datus est. Datus est ergo *EBA* trigonus specie. Proportio ergo recte *EB* ad *BA* data. Recte quoque *AB* ad *BG* proportio

1 XLIIII: 44[a] *B* / Si . . . . u]num *illeg. D* / angulum *tr. P post* habet / datum: datus *B* / alium *tr. B post* vero
2 ad[invicem proportionem *illeg. D*
3 trigonus *tr. B post* ille
4 alium. *tr. B post* vero
4–5 ad. . . .an]gulum *illeg. D*
5 angulum *om. B*
6–7 trigonus. . . .ad *illeg. D*
8 AG:BG *BP* / cathetus. . . .datus *illeg. D* / *post* Quoniam *scr. BOP* ergo / est[1] *om. B*
9–10 est[1]. . . .trigonus *illeg. D*
10 BAD:BDA *B* / *de* BAD *add. mg. O* per XL / *de* Proportio *add. mg. D* ex diffinitione / *supra* data *add. O* ex diffinitione / AB:BA *B*
11 data. . . .propor]tio *illeg. D* / *supra* data[2] *add. O* per VIII
12 *de* angulus *add. mg. D* per diffinitionem / *de* ergo[1] *add. mg. O* per XLIII / BDG[2]:BGD *B*
12–13 Datus[2]. . . .angulus[2] *illeg. D*
13 *supra* BGD angulus *add. O* per diffinitionem
14 *supra* ABG *add. O* per XL
14–15 ABG. . . .obtus]us *illeg. D*
16–17 a puncto. . . .angulus[1] *illeg. D*
17 *supra* deinceps *add. O* per IIII / BEA:BAE *B*
17–18 Est. . . .EBA[2] *illeg. D*
18 *supra* EBA trigonus *add. O* ex XL
19 *de* Proportio *add. mg. D* per 40 / recte: relique *B* / *supra* recte EB *add. O* ex diffinitione / BA: AB *BP*
19–1(p.s.) BA. . . .data[2] *illeg. D*

XLIV. If a triangle has one given angle and the sides including another angle have a given ratio to each other, that triangle is given in species.

Let the triangle *ABG* have one given angle at *A* and let the sides *AB* and *BG* including another angle *ABG* have a given ratio to each other. I say that the triangle *ABG* is given in species.

Let the angle at *A* be not a right angle but let it be, in the first place, an acute angle, and let a perpendicular *BD* be drawn from the point *B* to *AG*. Since the angle *BDA* is given and the angle at *A* is also given, hence the remaining angle at *B* is given. Therefore the triangle *BAD* is given in species. Therefore the ratio *BA*:*BD* is given. But the ratio *AB*:*BG* is given. Therefore the ratio *BD*:*BG* is also given. And the angle *BDG* is a right angle. Therefore the triangle *BDG* is given in species. Therefore the angle *BGD* is given. But the angle *BAG* is given. Therefore the remaining angle *ABG* is given. Therefore the triangle *ABG* is given in species.

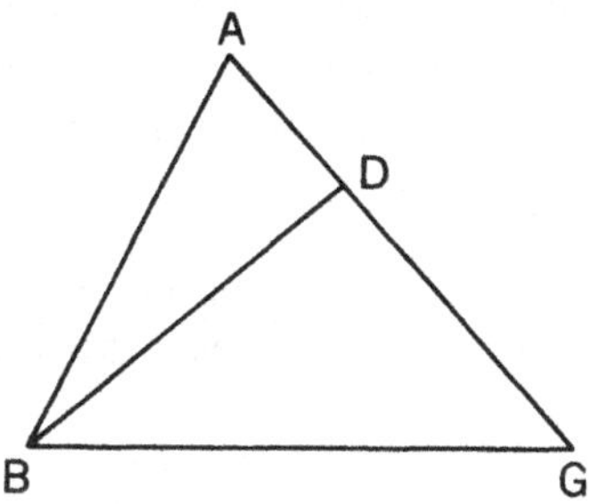

However, let the angle at *A* be obtuse and let the straight line *GA* be protracted to *E*, and let there be drawn from the point *B* to *AE* a perpendicular *BE*. Since the angle *BAG* is given, the adjacent angle *BAE* is also given. But the angle *BEA* is also given. Therefore the remaining angle *EBA* is also given. Therefore the triangle *EBA* is given in species. Therefore the ratio *EB*:*BA* is given. The ratio *AB*:*BG* is also given. Therefore

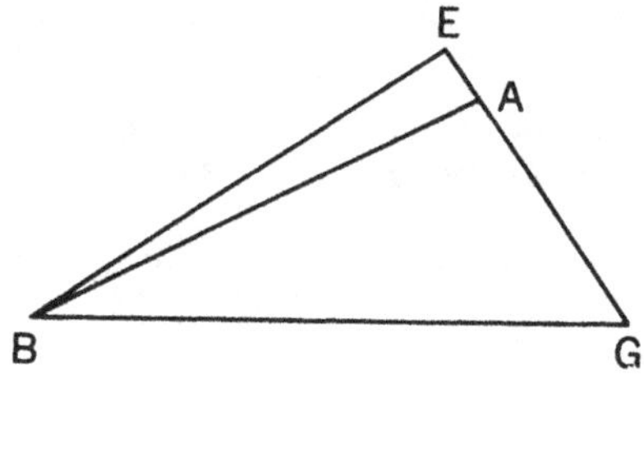

est data. Et *EB* ergo ad *BG* proportio est data. Et est rectus angulus *BEG*. Datus est ergo *EBG* trigonus specie. Datus ergo est angulus *BGE*. Estque et *BAG* angulus datus. Et reliquus ergo *ABG* angulus datus est. Datus est ergo *ABG* trigonus specie.

XLV. Si triangulus unum habet angulum datum, circa datum vero angulum latera coutraque sicut unum ad reliquum proportionem habent datam, datus est trigonus ille specie.

Esto trigonus *ABG* unum angulum datum habens *BAG*, circaque *BAG* latera, hoc est, coutrumque *BAG* sicut unum ad *GB* proportionem habeat datam. Dico quod *ABG* trigonus datus est specie.

Dividat *BAG* angulum in duo recta *AD*. Datus est ergo *BAD*. Et quoniam est sicut *BA* ad *AG* ita *BD* ad *DG*, permutatim sicut *AB* ad *BD* ita *AG* ad *GD*. Et sicut coutrumque ergo *BAG* ad *BG* ita *AB* ad *BD*. Proportio autem coutriusque *BAG* ad *BG* data. Proportio ergo et *BA* ad *BD*. data. Et est datus *BAD* angulus. Datus est ergo *ABD* trigonus specie. Datus ergo et *ABD* angulus. Est autem et *BAG* angulus datus. Et reliquus ergo *AGB* datus est. Datus est ergo *ABG* trigonus specie.

Aliter

Producatur *BA* recta super *D* lineeque *AG* equalis ponatur ad coiunganturque *D*, *G*. Et quoniam proportio est coutriusque *BAG* ad *GB* data,

1 *supra* EB *add. O* per VIII / est[3] *tr. D post* angulus
2 *supra* EBG *add. O* per XLIII / ergo[2] *tr. B post* est
2–3 Da[tus[1]....reliquus *illeg. D*
4 *de* Datus *add. mg. D* per 40 / est *tr. B post* ergo
5 XLV: 45[a] *B* / Si . . . . datum[1] *illeg. D* / habet *om. B*
5–6 circa . . . . angulum *om. B*
6 latera . . . . proportionem *illeg. D*
7–8 trigonus . . . . circa]que *illeg. D*
9 habeat: habet *P*
9–10 cou[trumque....quod *illeg. D*
10 specie *tr. B ante* datus
10–12 datus....sicut[1] *illeg. D*
11 BAG:BGD *B* / *supra* Datus *add. O* per II
12 *post* est *add. mg. O* per III VI[ti] elementorum
12–13 BD[1]....ita *illeg. D*
13 *supra* BAG ad BG *add. O* per XVI V[ti] elementorum
13–15 Proportio....BA]D *illeg. D*
15 est[2] *om. B* / *supra* specie *add. O* per premissam
15–16 Da[tus[1]....datus *illeg. D*
16–19 AGB....linee]que *illeg. D*
17 *post* ABG *add. O* ex XL
18–17(p.s.) Aliter....specie *om. B*
19–1(p.s.) ponatur....autem *illeg. D*

the ratio $EB:BG$ is given. The angle BEG is a right angle. Therefore the triangle EBG is given in species. Therefore the angle BGE is given. The angle BAG is also given. Therefore the remaining angle ABG is also given. Hence the triangle ABG is given in species.

XLV. If a triangle has one given angle and the line composed of the two sides including that given angle has a given ratio to the remaining side, that triangle is given in species.

Let there be a triangle ABG having one given angle BAG and let the sides including the angle BAG, that is, the sum of BA and AG, have a given ratio to GB. I say that the triangle ABG is given in species.

Let the angle BAG be bisected by a straight line AD. Therefore the angle BAD is given. Since $BA:AG = BD:DG$, by permutation, $AB:BD = AG:GD$. Therefore $(BA + AG):BG = AB:BD$. But the ratio $(BA + AG):BG$ is given. Therefore the ratio $BA:BD$ is given. And the angle BAD is given. Therefore the triangle ABD is given in species. Therefore the angle ABD is also given. But the angle BAG is also given. Therefore the remaining angle AGB is given. Therefore the triangle ABG is given in species.

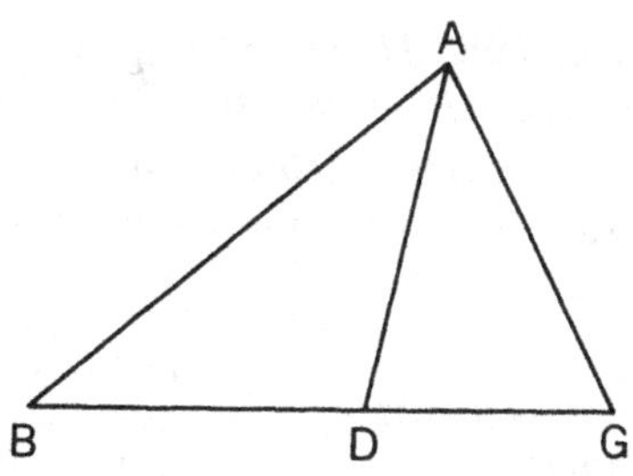

Another proof

Let a straight line BA be protracted to a point D, and let AD be equal to

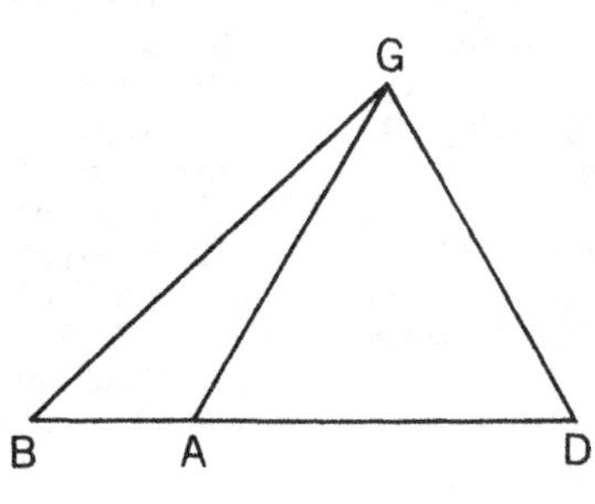

equalis autem *GA* recte *DA*, proportio ergo recte *BD* ad *BG* data. Estque datus *ADG* angulus, medietas enim est anguli *BAG*. Datus est ergo *BDG* trigonus specie. Datus ergo *ABG* angulus. Est autem et *BAG* angulus datus. Et reliquus ergo *AGB* datus est. Datus est ergo *ABG* specie.

XLVI. Si triangulus unum habet angulum datum, circa alium vero angulum latera coutraque sicut unum ad reliquum proportionem habent datam, datus est trigonus ille specie.

Esto trigonus *ABG* unum habens angulum datam *ABG*, circa alium vero angulum *BAG* latera coutraque sicut unum *BA*, *AG* ad *BG* proportionem (64v) habeant datam./Dico quod datus est trigonus ille specie.

Dividat ad *A* angulum in duo *AD* recta. Est ergo sicut coutrumque *BAG* ad *BG* ita *AG* ad *GD*. Proportio autem coutriusque *BAG* ad *GB* data. Proportio ergo et *AB* ad *BD* data. Et est datus qui ad *B* angulus. Datus est ergo *ABD* triangulus specie. Datus est ergo *BAD* angulus et est ei duplus *BAG*. Datus ergo est et *BAG* angulus. Est autem et *ABG* angulus datus. Et reliquus ergo qui ad *G* datus est. Datus est ergo *ABG* trigonus specie.

Aliter

Adiaceat recte *GA* equalis *AD* et coniungantur *D*, *G*. Quoniam ergo proportio est coutriusque *AB AG* ad *BG* data, equalis autem recta *GA* recte *AD*, proportio ergo et recte *DB* ad *GB* data. Et est datus qui ad *B* angulus. Datus est ergo *DBG* trigonus specie. Datus est ergo qui ad *D*

1–2 DA....BAG *illeg. D*
2 *post* Datus *add. mg. O* per XLIIII
2–3 BDG....BAG *illeg. D*
3 *supra* ABG angulus *add. O* per diffinitionem
4 *post* specie *add. O* per XL
4–6 datus[1]....an]gulum *illeg. D*
5–6 datum .... angulum *om. P*
6 coutr]aque *illeg. D*
6–7 reliquum .... da]tus *illeg. D*
7–9 ille .... BA *illeg. D*
10 habeant datam *illeg. D*
10–11 tri[gonus....du]o *illeg. D*
11 *post* ergo *add. mg. O* ex III VI[ti] elementorum et XVIII V[ti] elementorum ut in premissa
12 AG....ad[3] *illeg. D*
13–14 datus....triangulus *illeg. D*
14 *post* triangulus *add. mg. O* per XLI / *supra* BAD angulus *add. O* per diffinitionem
15 BAG[1]....ABG *illeg. D*
16 est[1] *om. DO* / *supra* trigonus *add. O* per XL
16–19 ergo[2]....Adiaceat *illeg. D*
20 ad BG *om. B* / equalis *tr. B post* autem
20–21 cou[triusque....da]tus *illeg. D*
21 ergo *om. B*
22 *de* ergo[1] *add. mg. O* per XLI / Da-[tus[2] *illeg. D* / *post* Datus[2] *bis (?) D* est ergo DBG trigonus specie / est[2] *om. B* / ergo[2] *tr. P ante* est[2]

AG and let *D* and *G* be joined. Since the ratio (*BA* + *AG*):*GB* is given and *DA* is equal to *GA*, hence the ratio *BD*:*BG* is given. And the angle *ADG* is given, for it is half of the [given] angle *BAG*. Therefore the triangle *BDG* is given in species. Therefore the angle *ABG* is given. But the angle *BAG* is also given. Therefore the remaining angle *AGB* is also given. Therefore [the triangle] *ABG* is given in species.

XLVI. If a triangle has one given angle and the line composed of the two sides including another angle has a given ratio to the remaining side, that triangle is given in species.

Let there be a triangle *ABG* having one given angle *ABG* and let the line composed of the two sides *BA* and *AG* about another angle *BAG* have a given ratio to *BG*. I say that that triangle is given in species.

Let the angle at *A* be bisected by a straight line *AD*. Therefore (*BA* + *AG*):*GB* = *AB*:*BD*. But the ratio (*BA* + *AG*):*GB* is given. Therefore the ratio *AB*:*BD* is also given. And the angle at *B* is given. Therefore the triangle *ABD* is given in species. Therefore the angle *BAD* is given and the angle *BAG* is double angle *BAD*. Therefore the angle *BAG* is also given. But the angle *ABG* is also given. Therefore the remaining angle at *G* is also given. Therefore the triangle *ABG* is given in species.

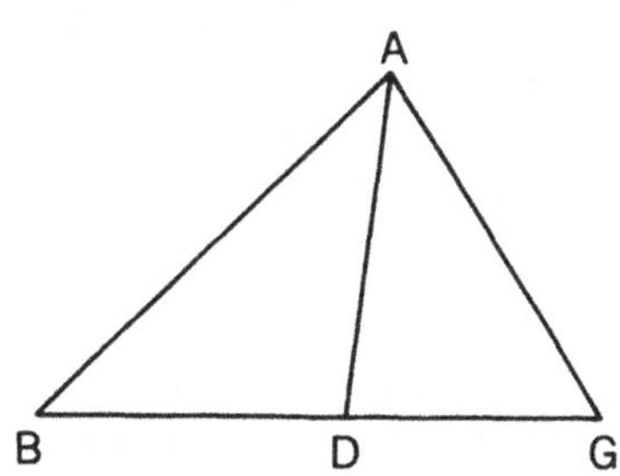

Another proof

Let *AD* be drawn equal to *GA* and let *D* and *G* be joined. Since the ratio (*AB* + *AG*):*BG* is given and *AD* = *GA*, hence the ratio *DB*:*BG* is also given. And the angle at *B* is given. Therefore the triangle *DBG* is given in species. Therefore the angle at *D* is given. And [angle] *BAG* is double it.

angulus. Et est ei duplus *BAG*. Angulus ergo *BAG* datus est. Datus est ergo *ABG* trigonus specie, quoniam et reliquus angulus datus est.

XLVII. Data rectilinea specie in data specie trigona dividuntur.

Esto datum rectilineum specie *ABGDE*. Dico quod *ABGDE* rectilineum in data specie trigona dividitur.

Coniungantur enim *BE* et *EG*. Quoniam ergo datum est *ABGDE* rectilineum specie, datus est ergo *BAE* angulus et est proportio *BA* ad *EA* data. Quoniam ergo datus est *BAE* angulus et est proportio *BA* ad *EA* data, datus est ergo *BAE* trigonus specie. Datus est ergo *ABE* angulus. Est autem et totus *ABG* angulus datus. Et reliquus ergo *EBG* datus est. Et est proportio *AB* ad *BE* data. Recte quoque *AB* ad *BG* proportio est data. Et *EB* ergo ad *BG* proportio est data. Et est datus *GBE* angulus. Datus est ergo *BGE* trigonus specie. Per eadem vero et *GDE* trigonus specie datus est. Data ergo rectilinea specie in data specie trigona dividuntur.

1 *post* angulus *add. mg. O* per diffinitionem / BAG: BG *D* / Angulus ergo BAG *om. B*
1–2 Datus. . . . et *illeg. D*
2 et *om. B* / *post* reliquus *add. mg. O* per XL
3 XLVII: 46[a] *B*
3–4 di[viduntur . . . . rectiline]um[1] *illeg. D*
4 datum *tr. B post* rectilineum[1] / rectilineum[1]: rectilinea *D* / Dico quod ABGDE *om. D*
6 Coniunga[ntur. . . . da]tum *illeg. D*
7 BAE: B *D* / *de* BAE *add. mg. D* per diffinitionem / *supra* BAE *add. O* ex diffinitione
7–8 BA. . . . ergo *illeg. D*
8 Quoniam. . . . data *om. P*
9 *supra* BAE *add. O* per XLI / trigo[nus. . . . ABE *illeg. D* / *de* BAE *add. mg. D* per 41
10 *de* datus[1] *add. mg. D* per 4 / Et[1] *om B* / *supra* EBG *add. O* per IIII
10–11 est. . . . BE *illeg. D*
12 Et. . . . data *om. D* / *supra* proportio *add. O* per VIII
12–13 Datus. . . . trigonus[1] *illeg. D*
13 et *om. P*
14 in. . . . divi]duntur *illeg. D*

Therefore the angle *BAG* is given. Therefore the triangle *ABG* is given in species, since the remaining angle is also given.*

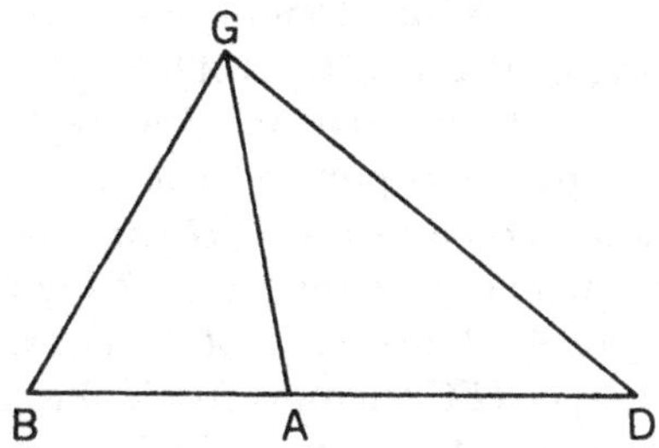

XLVII. Rectilineal figures given in species are divided into triangles given in species.

Let *ABGDE* be a rectilineal figure given in species. I say that the rectilineal figure *ABGDE* is divided into triangles given in species.

Let *B*, *E* and *E*, *G* be joined. Since the rectilineal figure *ABGDE* is given in species, the angle *BAE* is given and the ratio *BA*:*EA* is given. Since the angle *BAE* is given and the ratio *BA*:*EA* is given, hence the triangle *BAE* is given in species. Therefore the angle *ABE* is given. But the whole angle *ABG* is also given. Therefore the remaing angle *EBG* is also given. And the ratio *AB*:*BE* is given. The ratio *AB*:*BG* is also given. Therefore the ratio *EB*:*BG* is given. And the angle *GBE* is given. Therefore the triangle *BGE* is given in species. By the same argument, the triangle *GDE* is also given in species. Therefore the rectilineal figures given in species are divided into triangles given in species.

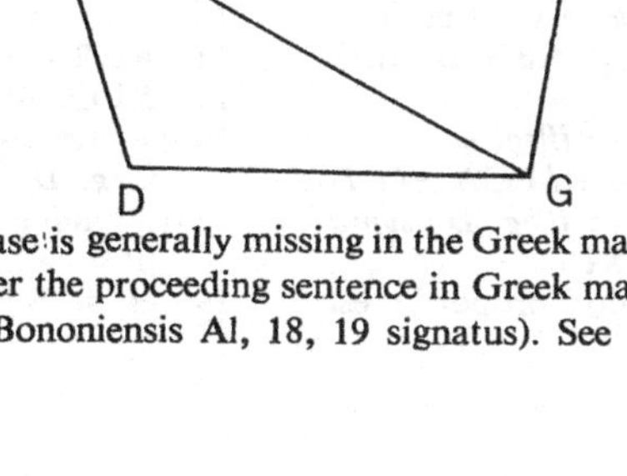

* The concluding phrase is generally missing in the Greek manuscripts but is present in an expanded form after the proceeding sentence in Greek manuscript b (= cod. bibliothecae communalis Bononiensis Al, 18, 19 signatus). See Menge, p. 202.

XLVIII. Si ab eadem recta duo trigoni describantur dati specie, proportionem habebunt adinvicem datam.

Ab eadem enim recta *AB* duo trigoni dati specie describantur *ABG*, *ADB*. Dico quod proportio est *ABG* ad *ADB* data.

Trahantur enim ab *A*, *B* punctis recte *AB* ad rectos *AE*, *IB* et educantur super puncta *Z*, *T* et per *G*, *D* puncta recte *AB* equidistantes trahantur *EGI*, *ZDT*. Quoniam ergo datus est *ABG* trigonus specie, proportio est *GA* ad *BA* data. Quoniam ergo datus est *GAB* angulus, est autem et *EAB* datus, et reliquus ergo *EAG* datus. Sed et *AEG* datus. Et reliquus ergo *EGA* datus est. Datus est ergo *AEG* trigonus specie. Proportio ergo *EA* ad *AG* data. Sed *AG* ad *AB* data est proportio. Et *EA* ergo ad *AB* proportio est data. Et per eadem etiam *ZA* ad *AB* proportio est data. Quare et *EA* ad *AZ* proportio est data. Et est sicut *AE* ad *AZ* ita *AI* ad *TA*. Quare et *AI* ad *AT* proportio est data. Et est quantitatis *AI* quidem medietas *ABG* et quantitatis *AT* medietas *ADB*. Et *ABG* ergo ad *ABD* proportio est data.

XLIX. Si ab eadem recta duo rectilinea quelibet describantur data specie, proportionem habebunt adinvicem datam.

Ab eadem enim recta *AB* duo rectilinea que continget data specie describantur *AEGZB* et *ADB*. Dico quod proportio est *AEGZB* ad *ADB* data.

1 XLVIII: 47[a] *B*
1–2 da[ti . . . . ha]bebunt *illeg. D*
3 trigoni. . . .descri]bantur *illeg. D*
4 quod. . . .ad *illeg. D*
5–6 ad. . . .AB *om. B* / ed[ucantur . . . . et *illeg. D*
7 dat[us. . . .specie *illeg. D*
8 *supra* GA ad *add. O* per diffinitionem / *de* data *add. mg. D* per diffinitionem / GAB: ABG *B* ABG trigonus *D*
8–9 est[2]. . . .re]liquus[1] *illeg. D*
9 et[1] *om. B* / et[2] *om. P* / EGA: EG *D*
10 Datus. . . .trigo]nus *illeg. D* / *supra* AEG *add. O* per XL
11 data est proportio: proportio est data *P*
11–12 Sed. . . .data[2] *om. B* / propor[tio[1]. . . .proportio *illeg. D*
12 *supra* data[1] *add. O* per VIII
13 *post* et *add. mg. O* per VIII / pro-[portio. . . .sicut *illeg. D* / *post* proportio *scr. B* ergo / est[2] *om. B* / AZ:EZ *B* / *de* ita *add. mg. D* per 1, 6 elementorum
14 Et. . . .medie]tas *illeg. D*
15 ADB:ABD *B* / Et. . . .data *illeg. D*
16 XLIX: 48[a] *B* / duo: 2 *B*
16–17 re[ctilinea . . . . propor]tionem *illeg. D*
17–19 ad[invicem . . . . AEGZB[2] *illeg. D*
19 et: ad *B* / quod *om. B* / ad *om. P*

XLVIII. If two triangles given in species are described on the same straight line, those triangles will have a given ratio to each other.

Let there be described on the same straight line *AB* two triangles given in species, *ABG* and *ADB*. I say that the ratio $\Delta ABG:\Delta ADB$ is given.

From the points *A* and *B* let there be drawn the lines *AE* and *IB* at right angles to the straight line *AB* and extended to the points *Z* and *T*, and through the points *G* and *D* let there be drawn *EGI* and *ZDT* parallel to the straight line *AB*. Since the triangle *ABG* is given in species, the ratio *GA*:*BA* is given. Since the angle *GAB* is also given and [the angle] *EAB* is also given, hence the remaining angle *EAG* is also given. But [the angle] *AEG* is also given. Therefore the remaining angle *EGA* is given. Therefore the triangle *AEG* is given in species. Therefore the ratio *EA*:*AG* is given. But the ratio *AG*:*AB* is given. Therefore the ratio *EA*:*AB* is given. And by the same argument, the ratio *ZA*:*AB* is also given. Hence the ratio *EA*:*AZ* is also given. And $AE:AZ = AI:TA$. Hence the ratio *AI*:*AT* is also given. And $\Delta ABG = 1/2\square AI$ and $\Delta ADB = 1/2\square AT$. Therefore the ratio ΔABG: ΔABD is given.

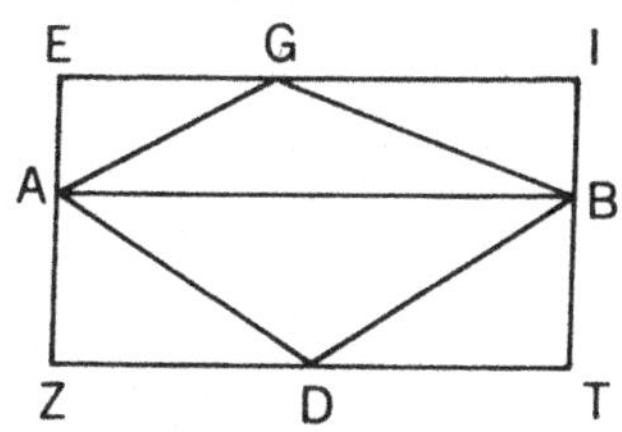

XLIX. If on the same straight line there are described any two rectilineal figures given in species, they will have a given ratio to each other.

On the same straight line *AB* let there be described any two rectilineal

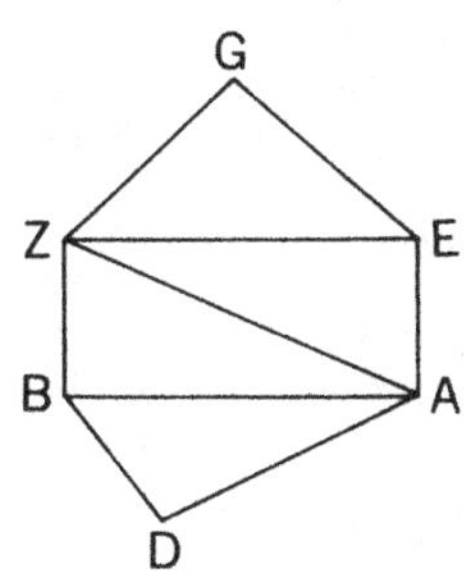

Producantur enim *AZ* et *ZE*. Datum ergo unumquodque horum *EGZ*, *EZA*, *ZAB* trigonum specie. Et quoniam ab eadem recta que est *EZ* duo trigona data specie descripta sunt que sunt *EZG* et *EZA*, proportio ergo est *GEZ* ad *ZEA* data. Et componenti ergo proportio est *GEAZ* ad *ZEA* data. Atqui *ZEA* ad *ZAB* proportio est data, quoniam quidem ab eadem recta que est *AZ* descripta sunt. Et *GEAZ* ergo ad *ZAB* proportio est data. Et componenti proportio *GEABZ* ad *ZBA* est data. Sed et *ZAB* ad *ADB* proportio est data. Et proportio ergo *GEABZ* ad *ADB* est data.

L. Si due recte adinvicem proportionem habent datam, et ab eis rectilinea et similia et similiter descripta adinvicem proportionem habebunt datam.

Due enim recte *AB*, *GD* adinvicem proportionem habebant datam et describantur ab *AB*, *GD* et similia et similiter iacentia rectilinea que sint *E*, *Z*. Dico quod et ea proportionem adinvicem habebunt datam.

Sumatur enim post *AB* et *GD* tertia proportionalis que sit *I*. Est ergo sicut *AB* ad *GD* ita *GD* ad *I*. Proportio autem recte *AB* ad *GD* data. Sicut vero *AB* ad *I* ita *E* ad *Z*. Proportio ergo et *E* ad *Z* data.

1 EGZ:EG *B*
1–2 AZ. . . .quoniam *illeg. D*
2 EZA:ZE *B* / *supra* trigonum *add. O* per XLVII / *post* eadem *add. mg. O* per premissam
2–3 du[o. . . .ergo *illeg. D*
3 descripta: protracta *B* / que *om. B* EZG:EZ *B*
4 *supra* componenti *add. O* per VI
4–5 componenti. . . .ZAB *illeg. D*
5–6 quidem. . . .propor]tio *illeg. D*
6 *post* ad *add. mg. O* per VIII
7 *supra* componenti *add. O* per VI / proportio: ergo *O*
7–8 compone[nti. . . .GEABZ *illeg. D*
8 Et *om. B* / GEABZ:GEABC *P*
9 L: 59[a] *B*
9–10 re[cte. . . .et[1] *illeg. D*
10 et[2] *om. B* / et[2]. . . .adinvicem *illeg. D*
12–13 habebant. . . .rectili]nea *illeg. D*
13–14 sint. . . .habebunt *illeg. D*
15 tertia *tr. B post* proportionalis
15–16 Sumatur. . . .GD[2] *illeg. D*
16–17 Pro[portio. . . .data *illeg. D*
17 AB:AD *B* / ergo *om. B*

figures given in species, *AEGZB* and *ADB*. I say that the ratio *AEGZB*:*ADB* is given.

Let *AZ* and *ZE* be drawn. Then each triangle* of *EGZ*, *EZA* and *ZAB* is given in species. Since on the same straight line, which is *EZ*, there are described the two triangles given in species, which are *EZG* and *EZA*, hence the ratio *GEZ*:*ZAE* is given. And by composing, the ratio *GEAZ*:*ZEA* is given. Now the ratio *ZEA*:*ZAB* is given, since they are described on the same straight line, which is *AZ*. Therefore the ratio *GEAZ*:*ZAB* is also given. And, by composing, *GEABZ*:*ZBA* is given.** But the ratio *ZAB*:*ADB* is also given. Therefore the ratio *GEABZ*:*ADB* is given.

L. If two straight lines have a given ratio to each other, similar rectilineal figures similarly described on those lines will also have a given ratio to each other.

Let two straight lines *AB* and *GD* have a given ratio to each other and on *AB* and *GD* let there be described similar rectilineal figures similarly placed, which are *E* and *Z*. I say that they will have a given ratio to each other.

To the two lines *AB* and *GD* let there be taken the third proportional which is *I*. Then *AB*:*GD* = *GD*:*I*. Now *AB*:*GD* is given.*** But *AB*:*I* = *E*:*Z*. Therefore *E*:*Z* is also given.

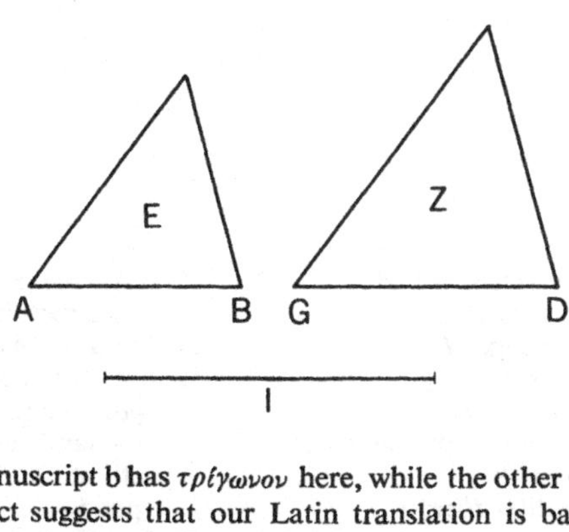

* Only the Greek manuscript b has τρίγωνον here, while the other Greek manuscripts have τριγώνων. This fact suggests that our Latin translation is based partially on a Greek text of the manuscript b. cf. Menge, p.90.

** The Latin translation agreeing with the Greek manuscript is again a manuscript of the 11th century at Bologna (= MS b) in having *GEABZ*, while the other Greek manuscripts merely have *GEBZA*.

*** Greek text adds here: "Therefore the ratio *GD*: *I* is also given. So the ratio *AB*: *I* is also given."

(65r) /LI. Si due recte adinvicem proportionem habent datam et ab eis rectilinea que contingent describantur data specie, proportionem habebunt ad invicem datam.

Due enim recte *AB*, *GD* adinvicem proportionem habeant datam et describantur ab eis rectilinea que contingent data specie que sint *E* et *Z*. Dico quod *E* ad *Z* proportio est data.

Describatur ab *AB* rectilineo *Z* simile et similiter iacens rectilineum *AIB*. Datum est autem *Z* specie. Datum est ergo et *AIB* specie. Sed et *E* datum est specie. Et descripta sunt ab eadem recta que est *AB*. Proportio ergo rectilinei *E* ad *AIB* data. Et quoniam proportio est recte *AB* ad *GD* data et descripta sunt ab eis similia et similiter iacentia *AIB* et *Z*, proportio ergo et rectilinei *AIB* ad *Z* data. Et rectilinei *AIB* ad *E* proportio est data. Et rectilinei ergo *E* ad *Z* proportio est data.

LII. Si a data recta magnitudine data specie figura describatur, data est ea descripta magnitudine.

A data enim recta magnitudine *AB* data specie figura describatur *AGDEB*. Dico quod *AGDEB* data est mangitudine.

1 LI *om. B* / adinvicem *om. B*
1–2 recte . . . . data *illeg. D*
2 contingent: continget *BP*
4 habeant: habent *P*
4–5 Due. . . .ab *illeg. D*
5 contingent *correxi ex* continget *BOP* / sint: sit *B*
5–7 que[1]. . . .similiter *illeg. D*
7 rectilineo *correxi ex* rectilineum *O* rectilinea *BP* / Z: et *P* / rectilineum: rectilinea *P*
7–9 rectilineum. . . .datum *illeg. D*
8 Datum[1]: Data *B* / est *om. B* / AIB[2]: AB *B*
9–10 ab. . . .Et *illeg. D*
10 ergo *om. P* / rectilinei: rectilinea *B* / *supra* rectilinei *add. O* per XLIX
11 data. . . .AIB *illeg. D* / et[3]:ad *B*
12 *de* rectilinei[1] *add. mg. D* per 50 / *post* rectilinei[1] *add. mg. O* per premissam / est *om. B* / rectilinei[2]: rectilinea *P* / de rectilinei[1] *add. mg. D* per 50
12–13 Z. . . .Et *illeg. D*
13 *post* ergo *add. mg. O* per VIII / *de* rectilinei *add. mg. D* per 8
14 LII: 51[a] *B*
14–15 magnitu[dine . . . . est *illeg. D*
17 est *om. B*

LI. If two straight lines have a given ratio to each other, and if on them there are described any rectilineal figures given in species, these figures will have a given ratio to each other.

Let the two straight lines *AB* and *GD* have a given ratio to each other and on them let there be drawn any rectilineal figures given in species, which we let be *E* and *Z*. I say that the ratio *E*:*Z* is given.

On *AB* let the rectilineal figure *AIB* be described, which is similar to, and similarly placed as, the rectilineal figure *Z*. Now *Z* is given in species. Therefore *AIB* is also given in species. But *E* is also given in species. And they have been described on the same straight line, which is *AB*. Therefore the ratio *E*:*AIB* is given. Since the ratio *AB*:*GD* is given, and on these lines *AIB* and *Z*, similar and similarly placed, have been described, hence the ratio *AIB*:*Z* is given. And the ratio *AIB*:*E* is given Therefore the ratio *E*:*Z* is also given.

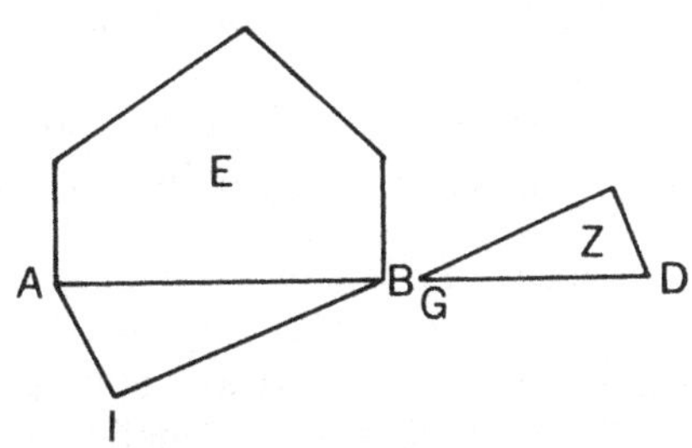

LII. If on a straight line given in magnitude there is described a figure given in species, that figure described is given in magnitude.

On a straight line *AB* given in magnitude, let there be described a figure *AGDEB* given in species. I say that *AGDEB* is given in magnitude.

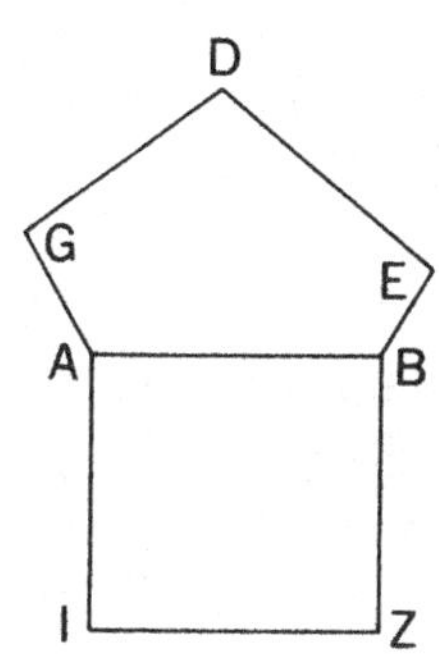

Describatur enim ab *AB* tetragonus *AZ*. Datum est ergo *AZ* specie et magnitudine. Et quoniam ab eadem linea *AB* duo rectilinea descripta sunt data specie *AGDEB* et *AZ*, proportio ergo rectilinei *AGDEB* ad *AZ* data. Sed datum est *AZ* magnitudine. Datum est ergo rectilineum *AGDEB* magnitudine.

LIII. Si due figure specie date fuerint et unum latus unius ad unum latus alterius proportionem habet datam, et reliqua latera ad reliqua latera proportionem habent datam.

Sint due figure specie date *AD* et *ET* et proportio lateris *BD* ad *ZT* data. Dico quod et reliquorum laterum ad reliqua latera proportio est data.

Quoniam enim proportio est *DB* ad *ZT* data, *DB* quoque ad *BA* proportio est data, et *AB* ergo ad *ZT* proportio est data. Atqui *ZT* ad *EZ* proportio est data. Et *AB* ergo ad *EZ* proportio eat data. Et per eadem quoque reliquorum laterum ad reliqua proportio est data.

LIIII. Si due figure date specie adinvicem proportionem habent datam et latera earum adinvicem proportionem habent datam.

Due enim figure date specie *A* et *B* adinvicem proportionem habeant datam. Dico quod et latera earum adinvicem proportionem habent datam.

Aut enim *A* ei quod est *B* simile est aut non. Esto primum simile. Sumaturque *GD* et *EZ* lineis tertia proportionalis *I*. Est ergo sicut *GD* ad *I* ita *A* ad *B*. Proportio autem *A* ad *B* data. Proportio ergo et *GD* ad *I* data.

1 Descri[batur. . . .est *illeg. D* / specie *om. P*
2–3 recti[linea. . . .AGDEB[1] *illeg. D*
3 *de* proportio *add. mg. D* per 49 / ergo *om. D* / *post* AGDEB[2] *add. mg.* O *per* XLIX
4 est[1] *om. DOP* / mag[nitudine. . . . rectilin]eum *illeg. D* / est[2] *om. B* / AGDEB:AGBED *B*
5 *post* magnitudine *add. mg. O* per II
6 LIII: 52[a] *B* / Si . . . . figure *illeg. D*
7 latus . . . . propor]tionem *illeg. D* / habet: habent *P* habeat *B*
7–8 ad . . . . latera *illeg. D*
8 latera *om. B* / propor[tionem . . . . datam *illeg. D*
10 quod et *illeg. D* / et *om. P*
11 DB[1]. . . .ad[2] *illeg. D* / *post* ZT *scr. P* proportio est *et del.* proportio
12 *de* data[1] *add. mg. D* per diffinitionem / ad[1] *om. D* / *post* ZT[1] *scr. et del. D* ad EZ / Atqui: Atque *B*
12–13 proportio[2]. . . .EZ *illeg. D*
15 LIIII: 53[a] *B* / Si due figure *illeg. D* / date *tr. B post* specie
16 la[tera . . . . adinvi]cem *illeg. D*
18 Dico. . . .adinvic em *illeg. D*
19–20 primum. . . .EZ *illeg. D*
20 ad *om. B*
21 A[1] *om. D* / Proportio[1]. . .Proportio[2] *illeg. D*

On *AB* let there be described the square *AZ*. Therefore *AZ* is given in species and magnitude. Since on the same line *AB* there are described the two rectilineal figures *AGDEB* and *AZ* given in species, therefore the ratio *AGDEB*:*AZ* is given. But *AZ* is given in magnitude. Therefore the rectilineal figure *AGDEB* is given in magnitude.

LIII. If there are two figures given in species and if one side of the one has a given ratio to one side of the other, the remaining sides [of the one] will also have a given ratio to the remaining sides [of the other].

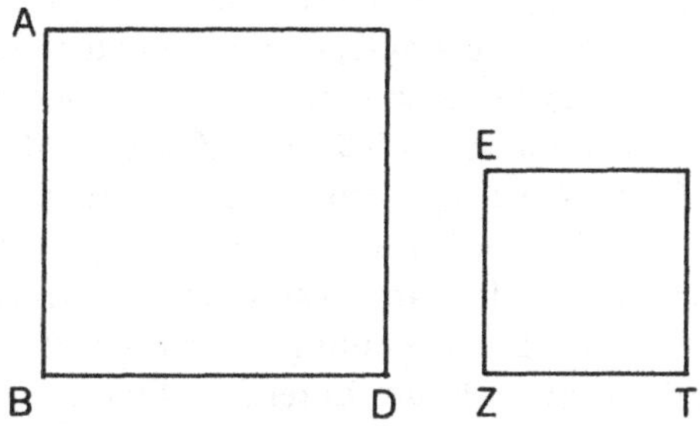

Let *AD* and *ET* be the two figures given in species, and let the ratio of side *BD* [to side] *ZT* be given. I say that the ratio of the remaining sides to the remaining sides is also given.

Since the ratio *DB*:*ZT* is given and *DB*:*BA* is also given, therefore the ratio *AB*:*ZT* is also given. But the ratio *ZT*:*EZ* is given. Therefore the ratio *AB*:*EZ* is also given. By the same argument, the ratio of the remaining sides to the remaining sides is also given.

LIV. If two figures given in species have a given ratio to each other, their sides also have a given ratio to each other.

Let two figures given in species, *A* and *B*, have a given ratio to each other. I say that their sides also have a given ratio to each other.

A is either similar to *B* or it is not. First let it be similar. And let there be taken the third proportional *I*, to the lines *GD* and *EZ*. Therefore *GD*:*I* = *A*:*B*. Now the ratio *A*:*B* is given. Therefore the ratio *GD*:*I* is given. But

Sunt autem *GD*, *EZ*, *I* proportionales. Et *GD* ergo proportio ad *EZ* data est. Et est simile *A* ei quod est *B*. Et reliqua ergo latera ad reliqua latera proportionem habent datam.

Non sit autem simile *A* ei quod est *B*, et describatur ab *EZ*, ei quod est *A* simile et similiter iacens *ET*. Datum est ergo et *ET* specie. Datum autem et *B*. Proportio ergo eius quod est *B* ad *ET* data. Eius quoque quod est *B* ad *A* proportio est data. Et *A* ergo ad *ET* proportio est data. Et simile *A* ei quod est *ET*. Proportio ergo *GD* ad *EZ* data. Et per eadem quoque reliquorum laterum ad reliqua latera proportio est data.

Aliter

Subiaceat data recta scilicet *IT*. Sed *A* ei quod est *B* aut simile est aut non. Esto primum simile, fiatque sicut *GD* ad *EZ* ita *IT* ad *KL*, et describantur ab *IT* et *KL* eis que sunt *A*, *B* similia et similiter iacentia *M*, *N*. Datum est ergo utrumque isotorum *M*, *N* specie. Et quoniam est sicut *GD* ad *EZ* ita *IT* ad *KL* et descripta sunt ab his *GD*, *EZ*, *IT*, *KL* similia et similiter iacentia rectilinea *A*, *B*, *M*, *N*, est ergo sicut *A* ad *B* ita *M* ad *N*. Proportio autem *A* ad *B* data. Proportio ergo *M* ad *N* data. Datum est autem *M*, a data enim recta magnitudine descripta est data figura. Datum ergo et *N*. Describatur autem ab *KL* tetragonum *Z*. Datum est ergo *Z* specie. Proportio ergo *N* ad *Z* data. Datum autem *N*. Datum ergo et *Z*. Data est autem *KL*. Est autem et *IT* data. Proportio ergo est *IT* ad *KL* data. Et est sicut *IT* ad *KL* ita *GD* ad *EZ*. Proportio ergo *GD* ad *EZ* data. Et est simile *A* ei quod est *B*. Et reliqua ergo latera ad reliqua latera proportionem habent datam. Non sit autem simile. Consequenter autem prime demonstrationi primi demonstrabitur.

1 ergo *tr. P post* proportio / *supra* ad EZ *add. mg. O* per XXIIII
1–2 ad. . . . B *illeg. D*
2 latera[2] *om. B*
4 Non. . . . ei[2] *illeg. D*
5 iacens. . . . et *om. B*
5–6 est. . . . B[2] *illeg. D*
6 ergo *tr. P post* eius / *supra* ET data *add. mg. O* per VIII
7 A[2] *om. B* / ET:BT *B* / proportio[1] . . . . A[3] *illeg. D*
8–9 da[ta. . . . proportio *illeg. D*
11–12 data. . . . primum *illeg. D*
12–13 ad. . . . simi]lia *illeg. D*
13 iacentia: iacencia *BP hic et ubique*
14 ergo: enim *D*
14–15 specie. . . . his *illeg. D*
15–16 similiter. . . . N[2] *illeg. D*
17 a: ad *P*
17–18 Proportio. . . . ma]gnitudine *illeg. D*
18 *supra* N *add. O* per II
18–19 data. . . . ergo *illeg. D*
19 est *om. P et tr. B post* ergo / Z[2]:GZ *B*
20 *supra* data *add. O* per XLIX / Datam autem N *om. P* / Datum[2]: Datus *B*
20–21 N[1]. . . . data[1] *illeg. D*
21–22 IT[2]. . . . Et *illeg. D*
22 Proportio. . . . EZ *om. B*
23 ad reliqua latera *om. B*
23–24 ei. . . . autem[1] *illeg. D*
24–25 Consequenter. . . demonstrabitur *illeg. D*

GD, *EZ* and *I* are proportionals. Therefore the ratio *GD*:*EZ* is given. And *A* is similar to *B*. Therefore the remaining sides also have a given ratio to the remaining sides.

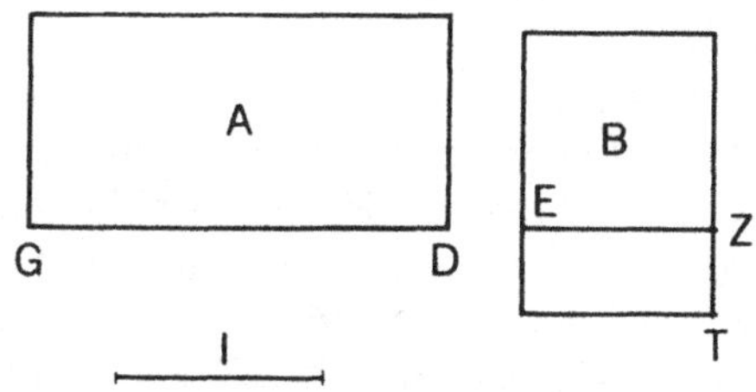

Let *A* be not similar to *B* and let there be described on *EZ* the figure *ET*, similar to, and similarly placed as, *A*. Therefore *ET* is also given in species. Now *B* is also given. Therefore the ratio *B*:*ET* is given. But the ratio *B*:*A* also is given. Therefore the ratio *A*:*ET* is given. And *A* is similar to *ET*. Therefore the ratio *GD*:*EZ* is given. By the same argument, the ratio of the remaining sides to the remaining sides is also given.

Another proof

Let the given straight line *IT* be posited. Now either *A* is similar to *B* or it is not. First, let it be similar, and let *GD*:*EZ* = *IT*:*KL*, and also on *IT* and *KL* let the figures *M* and *N* be described, similar to, and similarly placed as, *A* and *B*. Therefore each of *M* and *N* is given in species. But since *GD*:*EZ* = *IT*:*KL*, and on *GD*, *EZ*, *IT* and *KL* there are described the figures *A*, *B*, *M* and *N*, similar and similarly placed, therefore *A*:*B* = *M*:*N*. Now *A*:*B* is given. Therefore *M*:*N* is given. But *M* is given, for it is a figure given [in species] and described on a straight line given in magnitude. Therefore *N* is also given. Now let the square *Z* be described on *KL*. Then [figure] *Z* is given in species. Therefore the ratio *N*:*Z* is given. Now *N* is given. Therefore *Z* is also given. But *KL* is given. And *IT* is also given. Therefore the ratio *IT*:*KL* is given. And *IT*:*KL* = *GD*:*EZ*. Therefore the ratio *GD*:*EZ* is given. And *A* is similar to *B*. Therefore the remaining sides also have a given ratio to the remaining sides. Let *A* be not similar to *B*. But this case will be demonstrated as a consequence by the prior demonstration of the first [proof].

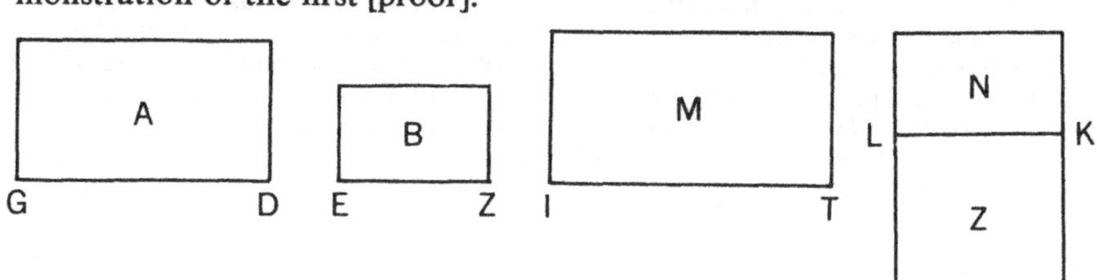

(65v) /LV. Si spatium specie et mangitudine datum fuerit, et latera ipsius magnitudine data erunt.

Esto spatium specie et magnitudine datum *A*. Dico quod et latera ipsius data sunt magnitudine.

Adiaceat enim positione et magnitudine data recta *BG* et describatur a linea *BG* ei quod est *A* simile et similiter iacens *D*. Datum est autem *D* specie. Et quoniam a data recta *BG* magnitudine data figura descripta est *D*, data est ergo et *D* magnitudine. Datum est autem et *A*. Proportio ergo *A* ad *D* data. Et est simile *A* ei quod est *D*. Proportio ergo *EZ* ad *BG* data. Data autem et *BG*. Data ergo et *EZ*. Et est proportio *ZE* ad *EI* data. Data ergo et *EI*. Et per eadem etiam unumquodque reliquorum datum est magnitudine.

Aliter

Esto spatium *KLMNZ* datum specie et magnitudine. Dico quod et latera eius data sunt magnitudine.

Describatur enim a linea *MN* tetragonum *MO*. Datum est ergo specie. Sed et *LN*. Proportio ergo *LN* ad *MO* data. Datum autem *LN* magnitudine. Datum ergo et *MO* magnitudine. Et est tetragonum a linea *MN*. Datum ergo quod a linea *MN*. Data ergo est recta *MN* magnitudine. Et per eadem etiam unumquodque horum *ML*, *LK*, *KZ*, *ZN* data ergo sunt magnitudine.

1 LV:54[a] *B* / et[1] *om. P*
1–2 Si . . . . ma]gnitudine *illeg. D*
2 magnitudine *tr. B post* erunt
2–4 magnitu[dine. . . . ma]gnitudine *illeg. D*
4–5 magni[tudine. . . . describa]tur *illeg. D*
6–7 li[nea. . . . recta *illeg. D*
7 *post* data[1] *scr. P* est / BG:AB *P*
7–8 magnitudine. . . . et[2] *illeg. D*
8 *post* magnitudine *add. mg. O* per LII
8–10 Propor[tio. . . . BG *illeg. D*
9 *supra* ad D *add. O* per L / *post* EZ *add. mg. O* per VI elementorum et XXIIII
10 *supra* EZ *add. O* per II
11–12 et. . . . magnitudine *illeg. D*
12 magnitudine *tr. B ante* datum
14 KLMNZ:KMLNZ *B*
14–15 KLM[NZ. . . . magnitudine *illeg. D*
16 Datum: Datus *B*
16–17 enim. . . . data *illeg. D*
17 *post* ergo LN *add. mg. O* per XLIX
17–18 mag[nitudine. . . . MN *illeg. D*
18 *post* MO magnitudine *add. mg. O* per II / tetragonum: tetragonus *P*
19 recta *om. B*
19–20 recta. . . . ML *illeg. D*

LV. If a space is given in species and magnitude, its sides will also be given in magnitude.

Let *A* be a space given in species and magnitude. I say that its sides are also given in magnitude.

Let the straight line *BG* given in position and magnitude be posited, and on the line *BG* let there be described *D*, similar to, and similarly placed as, *A*. Then *D* is given in species. And since *D* is a given figure described on the straight line *BG* given in magnitude, therefore *D* is also given in magnitude. Now *A* is also given [in magnitude]. Therefore the ratio *A*:*D* is given. But *A* is similar to *D*. Therefore the ratio *EZ*:*BG* is given. Now *BG* is also given. Therefore *EZ* is also given. And the ratio *ZE*:*EI* is given. Therefore *EI* is also given. By the same argument, each of the remaining sides is given in magnitude.

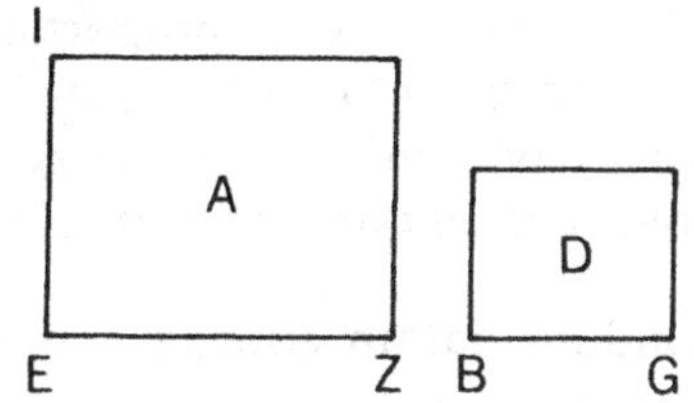

Another proof

Let the space *KLMNZ* be given in species and magnitude. I say that its sides are also given in magnitude.

On the line *MN* let the square *MO* be described. Therefore it is given in species. But also *LN* [is given]. Therefore the ratio *LN*:*MO* is given. Now *LN* is given in magnitude. Therefore *MO* is also given in magnitude. And the square is on the line *MN*. Therefore the square of line *MN* is given. Therefore the straight line *MN* is given in magnitude. By the same argument, therefore, each of *ML*, *KL*, *KZ* and *ZN* is also given in magnitude.

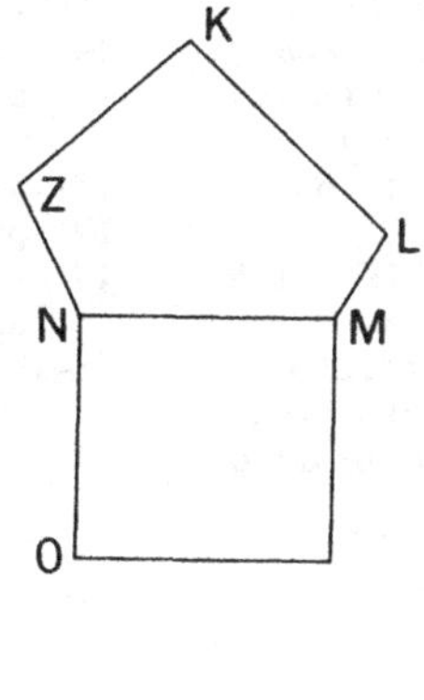

LVI. Si duo equiangula parallelogramma adinvicem proportionem habent datam, erit sicut primi latus ad secundi latus ita reliquum secundi latus ad quod alterum primi proportionem habent datam quam parallelogrammum ad parallelogrammum.

Duo enim equiangula parallelogramma *A*, *B* proportionem habeant datam. Dico quod est sicut *GD* ad *EZ* ita *EI* ad quod *GT* proportionem habet datam quam *A* parallelogrammum ad *B* parallelogrammum.

Eiciatur enim in directo recte *GT* recta *GK* et fiat sicut *GD* ad *EZ* ita *EI* ad *GK*, et compleatur *GL* parallelogrammum. Et quoniam est sicut *GD* ad *EZ* ita *EI* ad *GK*, equalis est autem *GD* ei que est *KL*, est ergo sicut *KL* ad *EZ* ita *EI* ad *GK*. Et circa equales angulos *GKL* et *IEZ* latera contrarie passa. Equalis ergo est *KD* ei quod est *IZ*. Et quoniam proportio est *A* ad *B* data, equalis autem *B* ei quod est *GL*, proportio ergo est *TD* ad *GL* data. Sicut autem *TD* ad *GL* ita *TG* ad *GK*. Et proportio ergo *TG* ad *GK* data. Et quoniam est sicut *GD* ad *EZ* ita *EI* ad *GK*, at vero *GT* ad *GK* proportionem habet datam quam spatium *A* ad *B*, est ergo sicut *GD* ad *EZ* ita *EI* ad quod *TG* proportionem habet quam spatium *A* ad spatium *B*.

LVII. Si datum secundum datam educatur in dato angulo, data est latitudo illius eductionis.

1 LVI: 55[a] *B* / parallelogramma *correxi ex* trigona *BDOP* / *supra* trigona *scr. manu rec. O* parallelogramma
1–2 adin[vicem . . . . erit *illeg. D*
2 reliquum: reliqui *P* secundum *B*
2–3 reliqu[um . . . . alterum *illeg. D*
4 parallelogrammum *correxi ex* parallellegramum *BO* paralelogramum *D* parallilogramum *P hic et ubique* / *supra* ad *scr. manu rec. O* habet / ad parallelogrammum *illeg. D*
5 equiangula: equianguli *P tr. B post* parallelogramma
6 quod. . . .quod *illeg. D* / GT:GI *B*
7 ad B parallelogramum *om. D*
8 GT. . . .EZ *illeg. D* / GT:TD *B*
10 ad[1]. . . .est[1] *illeg. D* / EI:GI *P* / GK: KG *B* EK *P* / est[1] *om. B* / que:quod *BP* / est[3] *om. B* / sicut *om. B*
11 GK:KG *B* / circa. . . .IEZ *illeg. D*
12 *de* Equalis *add. mg. manu rec. O* per XIIII 4[ti] euclidis / ergo est *om. P* KD:KDE *B* / ei *om. B* / est[2] *om. P* / proportio *tr. B post* est[3]
12–13 ad. . . .ei *illeg. D*
13 autem:A *B* / est[2] *om. B*
14 TG. . . .TG *illeg. D*
15 quoniam: quia *B* / GK[1]:TK *B* / GT:TG *B*
15–16 GK[2]. . . .quam *illeg. D*
17 propo[rtionem. . . .B *illeg. D*
18 LVII: 56[a] *B* / secundum *bis B* / datam: datus *B et corr. D ex* datum
18–19 angulo . . . . eductionis *illeg. D*

LVI. If two equiangular parallelograms have a given ratio to each other, one side of the first is to one side of the second, as the other side of the second is to the line to which the other side of the first has the given ratio which the one parallelogram has to the other parallelogram.

Let two equiangular parallelograms *A* and *B* have a given ratio. I say that *GD* is to *EZ* as *EI* is to the line to which *GT* has the given ratio which the parallelogram *A* has to the parallelogram *B*.

Let the straight line *GK* be produced in a straight line to form *GT* and let it be that $GD:EZ = EI:GK$, and let the parallelogram *GL* be completed. Since $GD:EZ = EI:GK$ and $GD = KL$, therefore $KL:EZ = EI:GK$. And the sides including the equal angles *GKL* and *IEZ* are reciprocally proportional. Therefore $KD = IZ$. Since the ratio $A:B$ is given and $B =$

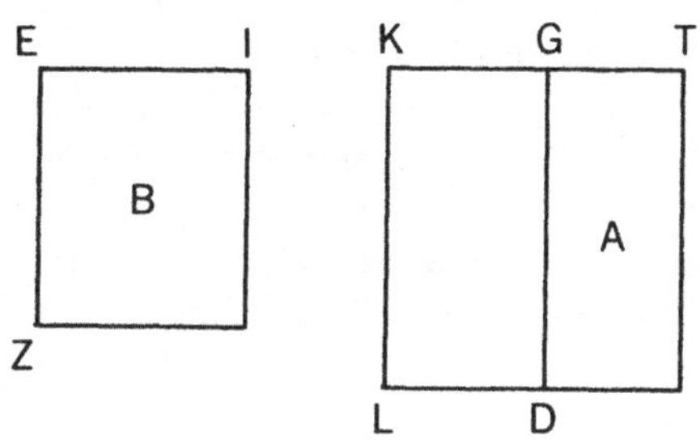

GL, therefore the ratio $TD:GL$ is given. Now $TD:GL = TG:GK$. Therefore the ratio $TG:GK$ is given. Since $GD:EZ = EI:GK$ and *GT* has to *GK* the given ratio which *A* has to *B*, therefore *GD* is to *EZ* as *EI* is to the line to which *TG* has the given ratio which the space *A* has to the space *B*.

LVII. If a given [space] is applied to a given [straight line] in a given angle, the width of that application is given.

Let a given [space] *AI* be applied to a given straight line *BA* in a given angle *GAB*. I say that *GA* is given.

Let the square *EB* be described on *AB*. Therefore *EB* is given. Let the lines *EA*. *ZB* and *GI* be protracted to the points *D* and *T*. And since each of *EB* and *AI* is given, the ratio $EB:AI$ is given. Now $AI = AT$. Therefore the ratio $EB:AT$ is given. Hence the ratio $EA:AD$ is given. Now $EA = AB$.

Datum enim *AI* secundum datam *BA* educatur in dato angulo *GAB*. Dico quod data est *GA*.

Describatur enim ab *AB* tetragonum *EB*. Datum est ergo *EB*. Protrahantur *EA*, *ZB*, *GI* super *D*, *T*. Et quoniam datum est utrumque horum *EB*, *AI*, proportio ergo *EB* ad *AI* data. Equalis autem *AI* ei quod est *AT*. Proportio ergo eius quod est *EB* ad *AT* data. Quare et *EA* ad *AD* proportio est data. Equalis autem *EA* ei que est *AB*. Proportio ergo *BA* ad *AD* data. Et quoniam datus est *GAB*, quorum *DAB* datus est, reliquus ergo *GAD* datus est. Est autem et *GDA* datus, rectus enim. Reliquus ergo *AGD* datus est. Datus est ergo *AGD* trigonus specie. Proportio ergo *GA* ad *AD* data. At vero *DA* ad *AB* proportio est data. Et *GA* ergo ad *AB* proportio est data. Et est data *BA*. Data ergo et *AG*. Et est latitudo eductionis.

LVIII. Si datum secundum datam educatur deficiens specie dato, date sunt latitudines defectionis.

Datum enim *AG* secundum datam *AD* educatur deficiens specie dato *DG*. Dico quod data est utraque *BG* et *BD*.

Dividatur enim *AD* in duo ad punctum *E*. Data est ergo *ED*. Et describatur ab *ED* ei quod est *GD* simile et similiter iacens rectilineum *EZ* et describatur secundum *EZ*. Datum est ergo *EZ* specie. Et quoniam a data recta *ED* data specie figura descripta est *EZ*, data est ergo *EZ* magnitudine. Et est ipsum equale eis que sunt *AG*, *KT*. Data sunt ergo *AG*, *KT* magni-

1–3 Dico. . . .Describatur *illeg. D*
3 tetragonum: tetragonus *P*
4 G[I. . . .utrumque *illeg. D* / est *om. B*
5 est *om. B*
5–6 EB. . . .ergo *om. D*
6 ergo *om. P* / Quare: Quoniam *B*
6–7 proportio. . . .AB *illeg. D*
7 que: quod *B* / *post* AB *bis D* proportio est data. Equalis autem EA ei que est AB / Proportio: Proportione *P* / BA:B *P*
7–8 data[2]. . . .quorum *illeg. D*
8 quoniam: quia *B* / DAB *tr. B post* datus[2] / est[2] *om. B* / ergo *om. B* / GAD:GAB *O*
9 *post* enim *scr. B* est
9–10 Est. . . .est[1] *om. D*
10 Datus. . . .AGD *om. B* / ergo *om. P*
10–11 Pro[portio. . . .vero *illeg. D*
11 proportio[1] *om. B* / est[1] *om. B*
12 est[1]. . . .eductionis *illeg. D*
13 LVIII: 57[a] *B* / deficiens: defficiens *B*
13–14 edu[catur . . . . sunt *illeg. D*
15 datam: lineam *B* / AD *om. B*
15–17 D[atum. . . .AD *illeg. D*
17 AD *om. D* / ergo *om. B*
17–18 est. . . .iacens *illeg. D*
18–19 describatur *tr. BP post* secundum
19–20 Datum. . . .specie *illeg. D*
20 *post* EZ[2] *bis B* specie. Et quoniam a data recta ED data specie figura est EZ, data est ergo EZ *et bis D* data est ergo EZ / *supra* magnitudine *add. O* per LII
20–1(p.s.) da[ta[2]. . . .mag]nitudine *illeg. D*
21 *post* ipsum *add. mg. O* per II elementorum / AG, KT: AGK *B*

Therefore the ratio $BA:AD$ is given. And since [the angle] GAB is given and its part $\angle DAB$ is given, therefore the remaining angle GAD is given. But the angle GDA is also given, for it is a right angle. Therefore the remaining angle AGD is given. Therefore the triangle AGD is given in species. Hence the ratio $GA:AD$ is given. But the ratio $DA:AB$ is given. Therefore the ratio $GA:AB$ is also given. And BA is given. Therefore AG is also given. And it is the width of the application.

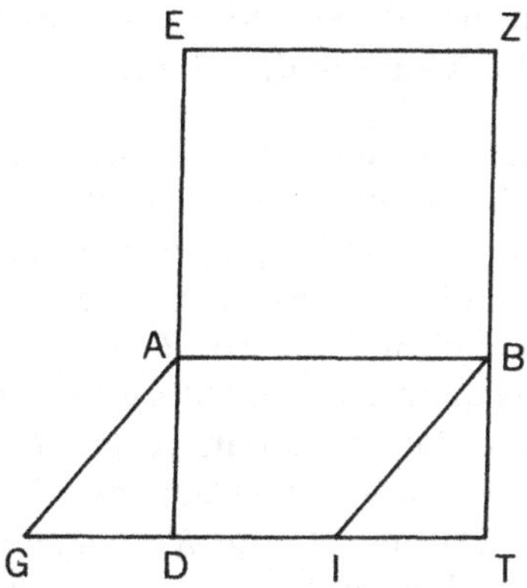

LVIII. If a given [space] is applied to a given straight line and it is deficient by a figure given in species, the sides of the deficient [space] are given.

Let a given [space] AG be applied to a given line AD, [the space] being deficient by a figure DG given in species. I say that each of BG and BD is given.

Let AD be divided into two [equal parts] at the point E. Therefore ED is given. And on ED let there be described the rectilinear figure EZ, similar to, and similarly placed as, GD, and let it be described according to the figure EZ. Therefore EZ is given in species. And since the figure EZ has been des-

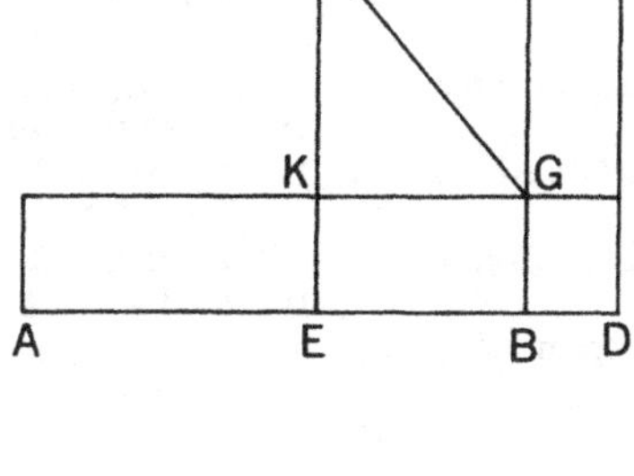

tudine. Et est *AG* datum magnitudine, suppositum est enim. Reliquum ergo *KT* datum est magnitudine. Est autem et specie datum, simile enim est ei quod est *GD*. Eius quod est *KT* ergo data sunt latera. Data ergo *GK*. Et est equalis ei que est *EB*. Data ergo est et *EB*. Est autem et *ED* data. Et reliquum ergo BD datum est. Et proportio *BD* ad *BG* data. Data ergo est et *BG*.

(66r) /LIX. Si datum secundum datam educatur superhabundans specie dato, date sunt latitudines superhabundantis.

Datum enim *AB* secundum datam *AG* educatur superhabundans specie dato *GB*. Dico quod date sunt uterque *TG* et *GE*.

Dividatur enim in duo equa *DE* ad punctum *Z* et describatur ab *EZ* ei quod est *BG* simile et similiter iacens *ZI*. Ergo circa eandem diametrum sunt *ZI* et *GB*. Trahatur eadem diametros *TEM* et describatur scema. Et quoniam simile est *GB* ei quod est *ZI*, datum est autem *GB* specie, datum est et *ZI* specie. Et descriptum est a data recta *ZE*. Datum est ergo *ZI* magnitudine. Est autem et *AB* datum. Data ergo sunt *AB* et *ZI*. Et sunt equalia ei quod est *KL*. Datum est ergo *KL* magnitudine. Est autem et specie, simile enim est *BG*. Figure ergo *KL* latera data sunt. Data ergo est *KT*. Quarum *KG* data est, equalis enim est *EZ*. Reliqua ergo *GT* data. Proportionem habet ad *TB* datam. Data ergo et *TB*.

1 est AG: DAG *B*
1–2 suppositum....magnitudine *om.* (?) *D*
1–3 magnitudine....Eius *illeg. D*
2 ergo *tr. B post* KT / enim est *om. B*
3–4 sunt....autem *illeg. D*
4 est[1] *tr. B post* equalis / que:quod *P* / EB:AB *B* / ergo *tr. B post* est[3]
5–6 reliqu[um....BG *illeg. D*
6 est *om. B*
7 LIX:58[a] B
7–8 secundum .... lati]tudines *illeg. D*
8 superhabundantis *correxi ex* educationis *BOP*
8–11 latitudin[es . . . . equa *illeg. D*
10 GE *corr. B ex* GD
11–12 pu[nctum....diame]trum *illeg. D*
13 sunt *om. B* / et[2]: ab B / describatur: ascribatur *O*
13–14 su[nt....quod *illeg. D*
14 datum[1]: data *B*
14–17 datum[2]....est[2] *illeg. D*
15 a: autem *B* / *de* Datum *add. mg. manu rec. O* per LII / ergo *om. B*
16 ergo *tr. B post* sunt / *de* Et *add. mg. manu rec. O* ex 36[a] primi libri(?)
17 ergo *om. B*
17–19 magnitudine . . . equalis *illeg. D*
18 ergo[1]: *tr. B post* est[2]
19 Quarum: Quare *BP* / Reliqua: Relique *BP* / GT:GA *B* / data[2] *correxi ex* datum *O* datam *BP*
19–20 EZ....TB[2] *illeg. D*

cribed on the given straight line *ED*, therefore *EZ* is given in magnitude. And $EZ = AG + KT$. Therefore $(AG + KT)$ is given in magnitude. And *AG* is given in magnitude, for it was so supposed. Therefore the remaining figure *KT* is also given in magnitude. But it [*KT*] is given also in species, for it is similar to *GD*. Therefore the sides of *KT* are given. Therefore *GK* is given. And $GK = EB$. Therefore *EB* is also given. Now *ED* is also given. Therefore the remainder *BD* is given. And the ratio $BD : BG$ is given. Therefore *BG* is also given.

LIX. If there is applied to a given straight line a given [space] which exceeds [it] by a figure given in species, the sides of the excess [space]* are given.

Let a given [space] *AB* be applied to a given straight line *AG*, [the space] exceeding [it] by a figure *BG* given in species. I say that *TG* and *GE* are given.

Let *DE* be divided into two equal parts at the point *Z*, and on *EZ* let the figure *ZI* be described, similar to, and similarly placed as, *BG*. Therefore *ZI* and *GB* are about the same diameter. Let the same diameter *TEM* be drawn and let that figure be described. And since *GB* is similar to *ZI* and *GB* is given in species, *ZI* is also given in species. And it is described on the given straight line *ZE*. Therefore *ZI* is given in magnitude. Now *AB* is also given. Therefore $(AB + ZI)$ is given. And $(AB + ZI) = KL$. Therefore *KL* is given in magnitude. But it is given also in species, for it is similar to *BG*. Therefore the sides of the figure *KL* are given. Therefore *KT* is given. Its part *KG* is given, for $KG = EZ$. Therefore the remaining line *GT* is given. It has a given ratio to *TB*. Therefore *TB* is also given.

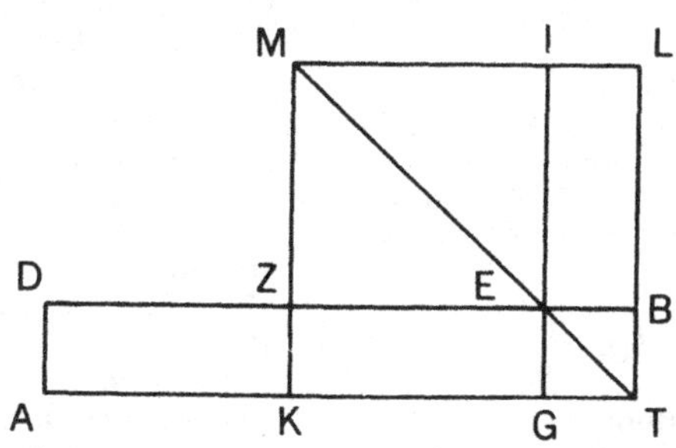

* The Latin has educationis. This is mistranslation, it ought to be superhabundantis.

LX. Si parallelogrammum datum specie et magnitudine dato gnomone augeatur vel minuatur, date sunt latitudines illius gnomonis.

Parallelogrammum enim *AB* datum specie et magnitudine augeatur dato gnomone *EGBDZI*. Dico quod data est utraque rectarum *GE* et *DZ*. Quoniam enim datum est *AB*, est autem et *EGBDZI* gnomon datus, et totum ergo *AI* datum est. Sed et specie, simile enim est ei quod est *AB*. Eius ergo quod est *AI* latera data sunt. Datum ergo utrumque horum scilicet *AE* et *AZ*. Est autem et utrumque horum videlicet *GA* et *AD* datum. Reliqua ergo utraque scilicet *EG* et *DZ* est data.

Rursum vero parallelogrammum *AI* datum specie et magnitudine minuatur dato gnomone *EGBDZI*. Dico quod data est utraque harum videlicet *GE* et *DZ*.

Quoniam enim datum est *AI*, cuius *EGBDZI* gnomon est datus, reliquum ergo *AB* datum est. Sed et specie. Eiusdem ergo reliqui AB latera sunt. Data est ergo utraque istarum scilicet *GA* et *AD*. Est autem et utraque harum *EA* videlicet et *AZ* data. Et reliqua ergo utraque scilicet *EG* et *DZ* data est.

LXI. Si date specie figure secundum unum laterum parallelogrammum spatium educatur in dato angulo, habet autem figura ad parallelogrammum proportionem datam, datum est parallelogrammum specie.

1 LX: 59[a] *B*
1–2 parallelogrammum . . . . latitudines *illeg. D*
2–3 gnomo[nis . . . . augea]tur *illeg. D*
3 *post* dato *scr. B* g
4 EGBDZI *correxi ex* EBDZI *BOP*
4–6 gnomo[ne. . . .totum *illeg. D*
5 datum: data *B* / EGBDZI *correxi ex* EBDZI *BOP* / gnomon: gnomo *B*
6 est. Sed *om. P*
6–7 Sed. . . .sunt *illeg. D*
7 latera *om. B*
7–8 scilicet. . . .datum *illeg. D*
9 est *correxi ex* sunt *OP et om. D*
9–10 data. . . .magnitudine *illeg. D*
10–11 minuatur: minnatur *BD*
11 EGBDZI: EGDZI *D* / utraque *tr. P ante* data / harum: istarum *B et om. P*
11–13 Dico. . . .d]atum *illeg. D*
13 EGBDZI:EGDBZI *D* / gnomon: gnomo *B* / est[2] *tr. B post* datus / reliquum: reliquam *O* relique *B*
13–14 datus. . . .Eius]dem *illeg. D*
14 est *om. B*
15–16 ergo. . . .harum *illeg. D* / harum: earum *B*
16 AZ:AT *B*
17 est *correxi ex* sunt *BDOP*
18 LXI: 60[a] *B* / Si . . . . secundum *illeg. D*
19 spatium educatur *illeg. D*
20 parallel[ogrammum proportione]m *illeg. D*
20–21 proportionem . . . . specie: spacium educatur *P*

LX. If a parallelogram given in species and magnitude is augmented or diminished by a give gnomon, the latitudes [*i.e.* determinate dimensions] of that gnomon are given.

Let the parallelogram *AB* given in species and magnitude be augmented by the given gnomon *EGBDZI*.* I say that each of the straight lines *GE* and *DZ* is given.

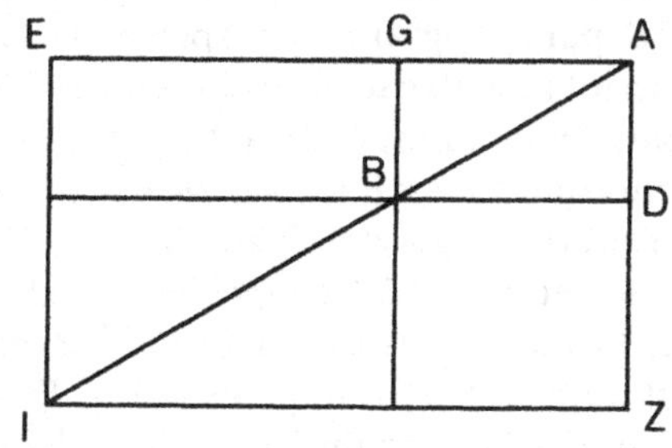

Since *AB* is given and the gnomon *EGBDZI* is given, the whole [parallelogram] *A* is also given. But it is also given in species, for it is similar to *AB*. Therefore the sides of *AI* are given. Therefore each of *AE* and *AZ* is given. Now each of *GA* and *AD* is given. Therefore each remaining line of *EG* and *DZ* is also given.

Further, let the parallelogram *AI* given in species and magnitude be diminished by a given gnomon *EGBDZI*. I say that each of *GE* and *DZ* is given.

Since *AI* is given and its gnomon *EGBDZI* is given, the remaining [parallelogram] *AB* is given [in magnitude], and [it is given] also in species. Therefore the sides of the same remaining [parallelogram] *AB* are given. Therefore each of *GA* and *AD* is given. Now each of *EA* and *AZ* is given. Therefore each remaining line of *EG* and *DZ* is also given.

LXI. If to one side of a figure given in species there is applied a parallelogrammic space in a given angle, and if the given figure has a given ratio to the parallelogram, the parallelogram is given in species.

* The original reading in Greek manuscript v was like the original Latin reading: *EBDZI*. See Menge, p. 108.

Date enim specie figure *AZGB* secundum unum laterum scilicet *GB* parallelogrammum spatium educatur *GD* in dato angulo *LGB*, proportio vero esto figure *AG* ad *GD* parallelogrammum data. Dico quod datum est parallelogrammum *GD* specie.

Trahatur enim per punctum quidem *B* recte *ZG* parallelos *BI*, per *Z* vero recte *GB* parallelos *ZI* et protrahatur *ZG* et *IB* super *T* et *K* puncta. Quoniam ergo datus est angulus *ZGB* et est proportio *ZG* ad *GB* data, datum est ergo *ZB* parallelogrammum specie. Data est autem specie *AZGB* figure et descripta est ab eadem recta scilicet *GB*. Proportio ergo et *AB* figure ad *ZB* parallelogrammum data. Sed et *ZB* ad *GD* proportio est data, quoniam et eius quod est *AB* ad *GD* supposita est. Equale vero *GD* ei quod est *KB*. Proportio ergo et *KB* ad *GI* data. Quare et *ZG* ad *GK* proportio est data. Sed et *ZG* ad *GB* proportio est data. Et *GB* ergo ad *GK* proportio est data. Et quia datus est angulus *ZGB*, et qui deinceps ergo *BGK* datus est. Est autem et *BGL* angulus datus. Et reliquus ergo *LGK* datus est. Est autem et *LKG* angulus datus, equalis enim angulo *KGB*. Reliquus ergo *GLK* datus est. Datus est ergo *LGK* trigonus specie. Proportio ergo est *LG* ad *GK* data. Sed et *KG* ad *GB* proportio est data. Et *LG* ergo ad *GB* proportio est data. Et est datus angulus *LGB*. Datum est ergo *GD* parallelogrammum specie.

LXII. Si due recte adinvicem proportionem habent datam, et

1 specie *tr. DO post* figure / AZGB: AGBZ *D* / secundum....GB *om. B*
2 parallelogrammum....in *illeg. D* / LGB: AGB *BP*
3 esto: est *B et bis P*
3–4 da[ta....parallelogrammum *illeg. D*
5 B *tr. P post* recte
6 vero....pro]trahatur *illeg. D* / parallelos *correxi ex* paralellos *BDO et* parallilos *P hic et ubique* / ZG: GZ *B*
7 ZGB....data *illeg. D* / ZGB:ZBG *B*
8 est[1] *bis P* / specie[2] *om. P*
9–10 de[scripta....scilicet *illeg. D* / *de* et AB *add. mg. O* per XLIX
10 et *om. B*
10–11 proportio....eius *illeg. D*
12 Propor[tio....Quar]e *illeg. D*
13 est[1] *om. B* / proportio est[2] *om. B* / GB[2]:BG *BP*
13–14 Et....quia *illeg. D*
14 *post* Et *bis P* GB ergo ad GK proportio est data. Et
15 BGK:GBK *D* / autem....reli]quus *illeg. D*
16 LKG:LGK *D*
16–17 angulo....da]tus *illeg. D*
17 GLK *corr. B ex* GKLK / Datus est *om. B* / LGK:LKG *BOP*
18 data....data *illeg. D*
19 ergo: igitur *B*
21 LXII: 61[a] *B* / habent: habeant *B*

To one side GB of the figure $AZGB$* given in species, let the parallelogrammic space GD be applied to a given angle LGB, and let the ratio of the figure AG to the parallelogram GD be given. I say that the parallelogram GD is given in species.

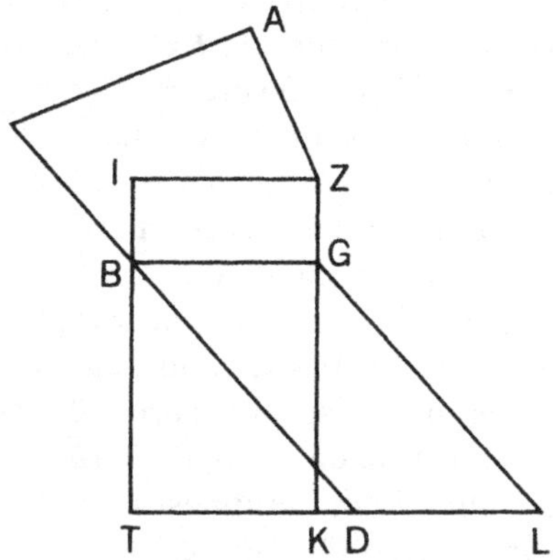

Through the point B, let BI be drawn parallel to the straight line ZG and through the point Z let ZI be drawn parallel to the straight line GB, and let ZG and IB be protracted to the points T and K. Since the angle ZGB is given and the ratio $ZG:GB$ is given, therefore the parallelogram ZB is given in species. Now the figure $AZGB$ is given in species and described on the same straight line GB. Therefore the ratio of the figure AB to the parallelogram ZB is also given. But the ratio $ZB:GD$ is also given, since $AB:GD$ was assumed as given. Now $GD = KB$. Therefore the ratio $KB:GI$ is given. Hence the ratio $ZG:GK$ is also given. But the ratio $ZG:GB$ is also given. Therefore the ratio $GB:GK$ is given. And since the angle ZGB is given, the adjacent angle BGK is also given. Now the angle BGL is given. Therefore the remaining angle LGK is also given. But the angle LKG is also given, for $\angle LKG = \angle KGB$. Therefore the remaining angle GLK is given. Therefore the triangle LGK is given in species. Therefore the ratio $LG:GK$ is given. But the ratio $KG:GB$ is also given. Therefore the ratio $LG:GB$ is given. And the angle LGB is given. Therefore the parallelogram GD is given in species.

LXII. If two straight lines have a given ratio to each other, and if on one of them there is described a figure given in species but on the

* Of the Greek manuscripts, MS Vat. has $AZGB$. See Menge, p. 112.

describatur ab una quidem data specie figura ab altera vero spatium parallelogrammum in dato angulo, habet autem figura ad parallelogrammum proportionem datam, datum est parallelogrammum specie.

Due enim recte *AB* et *GD* adinvicem proportionem habeant datam et describatur ab *AB* data specie figura *AEB* et a recta *GD* parallelogrammum *DZ* in dato angulo *ZGD*, proportio vero esto figure *AEB* ad *ZD* parallelogrammum data. Dico quod datum est *DZ* parallelogrammum specie. Describatur enim ab *AB* ei quod est *DZ* simile et similiter iacens paral- (66v) lelogrammum *AI*. Et quoniam proportio est *AB* ad *GD* data,/et descripta sunt ab *AB* et *GD* similia et similiter iacentia rectilinea videlicet *AI* et *ZD*, proportio est *AI* ad *ZD* data. Sed et *ZD* ad *EB* proportio est data. Et *EB* ergo ad *AI* proportio est data. Et est datus angulus *ATI*, equalis enim angulo *ZGD*. Quoniam ergo date specie figure *EB* secundum unum laterum videlicet *AB* eductum est *AI* in dato angulo *TAB* et proportio est *EB* figure ad *AI* parallelogrammum data, datum est ergo *AI* specie. Et simile est ei quod est *ZD*. Datum est ergo *ZD* specie.

LXIII. Si trigonus specie datus fuerit, cuiusque laterum eius tetragonum ad trigonum eundem proportionem habet datam.

Esto trigonus specie datus *ABG* et describantur a singulis ipsius lateribus tetragona *EB*, *GD*, *GZ*. Dico quod unumquodque eorum videlicet *EB*, *GD*, *GZ* ad *ABG* trigonum proportionem habet datam.

1 de[scribatur . . . . quidem *illeg. D* / spacium *tr. B ante* in *ln. 2*
1–2 spa[tium . . . . a]utem *illeg. D*
3 proportionem . . . . specie *illeg. D*
4 propor]tionem *illeg. D* / haneant: habent *P*
5–6 ab. . . .angu]lo *illeg. D*
6–8 figure. . . .De]scribatur *illeg. D*
8 AB:A *D* / DZ:EZ *B*
8–9 est. . . .ad *illeg. D*
9 est: etiam *B*
10 iacentia: iacencia *BP* / rectilinea: recta linea *B et tr. P post* videlicet
10–11 ab. . . .est[1] *illeg. D*
11 proportio. . . .ZD[2] *om. P*
11–12 et. . . .an]gulus *illeg. D*
12 est[1]: etiam *B* / est[2] *om. B*
13 specie *tr. P post* figure
13–14 ZGD. . . .in *illeg. D*
14–15 et. . . .est[1] *illeg. D*
15 est[1]: etiam *B* / *de* datum *add. mg. O* per premissam
16 est[1]: et *B*
17 LXIII:62[a] *B*
17–18 Si . . . . datam *illeg. D*
18 eundem *om. B*
19 Esto. . . .descri]bantur *illeg. D*
19–20 a. . . .tetragona *illeg. D*
20–21 GZ. . . .datam *illeg. D*

other a parallelogrammic space in a given angle, and if the figure has a given ratio to the parallelogram, the parallelogram is given in species.

Let the two straight lines *AB* and *GD* have a given ratio to each other, and on *AB* let there be described the figure *AEB* given in species and on *GD* the parallelogram *DZ* in a given angle *ZGD*, and let the ratio of the figure *AEB** to the parallelogram *ZD* be given. I say that the parallelogram *DZ* is given in species.

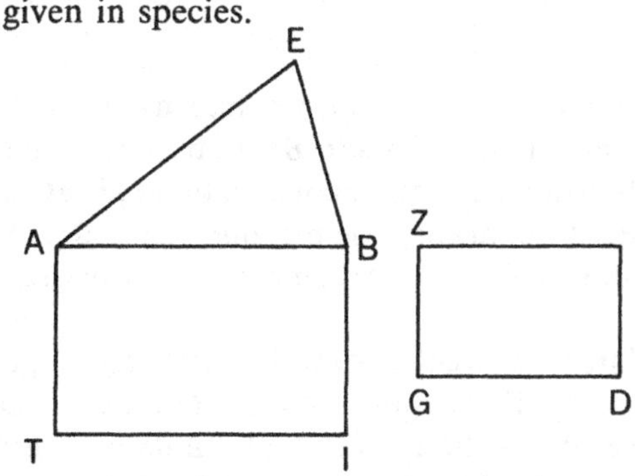

On the straight line *AB* let there be described the parallelogram *AI* similar to, and similarly placed as, *DZ*. Since the ratio *AB*:*GD* is given and on *AB* and *GD* there are described *AI* and *ZD*, similar and similarly placed, the ratio *AI*:*ZD* is given. But also the ratio *ZD*:*EB* is given. Therefore the ratio *EB*:*AI* is given. And the angle *ATI*** is given, for $\angle ATI = \angle ZGD$. Since to one of the sides of the figure *EB* given in species, namely to *AB*, the parallelogram *AI* is applied in the given angle *TAB* and the ratio of the figure *EB* to the parallelogram *AI* is given, therefore *AI* is given in species. And *AI* is similar to *ZD*. Therefore *ZD* is given in species.

LXIII. If a triangle is given in species, the square on each of its sides has a given ratio to the same triangle.

Let the triangle *ABG* be given in species and on each of its sides let the squares *EB*, *GD* and *GZ* be described. I say that each of *EB*, *GD* and *GZ* have a given ratio to the triangle *ABG*.

* Menge gives the Greek reading *EB* without noting a variant, but Peyrard gives *AEB*, as does our Latin translation. Perhaps Menge was in error here. Note also that Menge given figure *AE* not as a triangle but as an irregular quadrilateral. See Menge, p. 115.

** Menge gives the Greek reading *BAΘ* without variant. Peyrard gives it as *ABH*.

Quoniam enim ab eadem recta scilicet *BG* rectilinea data specie descripta sunt que contingit videlicet *ABG* et *GD*, proportio ergo *ABG* ad *GD* data. Et per eadem etiam utrumque istorum videlicet *EB* et *ZG* ad *ABG* trigonum proportionem habet datam.

LXIIII. Si trigonus obtusum habeat angulum datum, quo maius potest obtuso angulo subtensum latus eundem obtusum angulum continentibus laterbus, illud spatium ad trigonum proportionem habebit datam.

Esto trigonus ambligonius *ABG* obtusum habens angulum *ABG* datum et pertrahatur in directo linee *GB* recta *BD* et trahatur a puncto *A* super *GD* cathetus *AD*. Dico quoniam quo maius est quod ab *AG* quadratum eis que ab *AB* et *BG* fuerit quadratis, hoc est, quod bis sub *DB* et *BG* continetur rectangulum illud spatium ad trigonum *ABG* proportionem habebit datam.

Quoniam enim datus est angulus *ABG*, et *ABD* ergo angulus datus est. Est autem et *ADB* datus. Et reliquus ergo *DAB* datus est. Datus est ergo *DAB* trigonus specie. Proportio ergo *AD* ad *DB* data. Et est sicut *AD* ad *DB* ita quod sub *AD* et *BG* ad illud quod sub *DB* et *BG*. Quare et eius quod

1 *post* BG *scr. B* G
1–2 ab. . . .ad *illeg. D*
2 *post* ergo *scr. OP* est / ABG:AGB *B* / *post* data *add. mg. O* per XLLX
2–4 da[ta. . . .datam *illeg. D*
5 LXIIII: 63^a *B* / maius: magnum *B*
5–6 Si . . . . ob]tuso *illeg. D*
6–7 angulo. . . . contine]ntibus *illeg. D*
7–9 spatium . . . . an]gulum *illeg. D*
8 habebit: habet *B*
9 ambligonius: ambligonum *B*
10–11 per[trahatur. . . .GD *illeg. D*
11–12 maius. . . .qua]dratis *illeg. D*
12 que:quod *B* / DB:GB *B* / BG:BD *B*
12–13 conti[netur. . . .proportionem *illeg. D*
15 est^2: etiam *B*
15–16 ABD. . . .Et *illeg. D*
16 est^1: etiam *B*
17 specie. . . .sicut *illeg. D* / est: etiam *B* / *de* AD ad^2 *add. mg. manu rec. O* per primam 6^{ti} euclidis
18 *post* illud quod *add. mg. O* ex prima VI^{ti} elementorum / et^2: ad *B*
18–2(p.s) D[B^1. . . .BG^1 *illeg. D*
18–1(p.s) Quare. . . .BG^2 *om.* (?) *D*

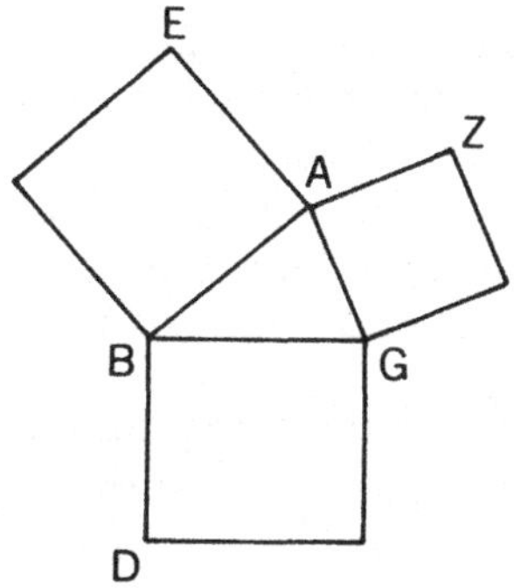

Since on the same straight line *BG* there are described any rectilineal figure given in species, which we let be *ABG* and *GD*, the ratio *ABG*:*GD* is given. By the same argument, each of *EB* and *ZG* have a given ratio to the triangle *ABG*.

LXIV. If a triangle has a given obtuse angle, that space by which the square of the side subtending the obtuse angle is greater than the squares of the sides containing the same obtuse angle will have a given ratio to the triangle.

Let the obtuse-angle triangle *ABG* have a given obtuse angle *ABG*, and let the straight line *BD* be drawn as a direct extension of *GB*, and from the point *A* let there be drawn the perpendicular *AD* to *GD*. I say that the space by which the square of *AG* is greater than the squares of *AB* and *BG*, that is, the double of the rectangle contained by *DB* and *BG* [*i.e.* 2 *DB* × *BG*], will have a given ratio to the triangle *ABG*.

Since the angle *ABG* is given, the angle *ABD* is also given. But the angle *ADB* is also given. Therefore the remaining angle *DAB* is given. Therefore the triangle *DAB* is given in species. Therefore the ratio *AD*:*DB* is given.

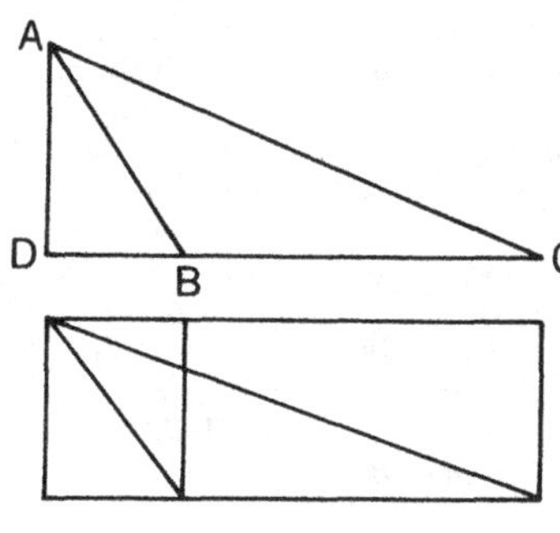

sub *DA* et *BG* ad illud quod sub *DB* et *BG* proportio est data. Et eius ergo quod sub *DB* et *BG* bis ad illud quod sub *AD* et *BG* proportio est data. Sed eius quod sub *DA* et *BG* ad *ABG* proportio est data. Et eius ergo quod sub *DBG* bis ad *ABG* trigonum proportio est data. Et est quod bis sub *DB* et *BG*, quo maius est quod ab *AG* quadratum eis que ab *AB* et *BG* quadratis. Illud ergo spatium ad *ABG* trigonum proportionem habebit datam.

LXV. Si trigonus acutum habeat angulum datum, quo minus potest acutum angulum subtendens latus lateribus acutum angulum continentibus, illud spatium ad trigonum proportionem habet datam.

Esto trigonus oxigonius *ABG* acutum habens angulum datum *ABG* et trahatur ab *A* puncto super *BG* cathetus *AD*. Dico quia quo minus est quod ab *AG* quadratum eis que ab *AB* et *BG*, hoc est, quod bis sub *GB*, *BD* ad *ABG* trigonum proportionem habet datam.

Quoniam enim datus est *ABD* angulus, est autem et *ADB* datus, et reliquus ergo *BAD* datus est. Datus est ergo *ABD* trigonus specie. Proportio ergo *BD* ad *DA* data. Quare et eius quod sub *GBD* ad illud quod sub *GB* et *AD* proportio est data. Et eius ergo quod bis sub *GB* et *BD*. Sed eius quod sub *BG* et *AD* ad *ABG* proportio est data. Et eius ergo quod bis sub *GB*, *BD* ad *ABG* trigonum proportio est data. Et est quod bis sub *GB*, *BD* quo minus est quod ab *AG* quadratum eis que ab *AB* et *BG*. Quo ergo minus est quod ab *AG* eis que ab *AB* et *BG*, illud spatium ad *ABG* trigonum proportionem habet datam.

1 est: etiam *B* / ergo *om. P*
1–4 Et. . . .data *om. B*
4 *de* est[2] quod *add. mg. manu rec. O* per XII 2[di] libri euclidis
4–5 propor[tio. . . .BG[1] *illeg. D*
5 quo: quod *B* / quod: quam *B* / que: quod *B*
5–6 quadratis. . . .datam *illeg. D*
7 LXV: 64[a] *B* / a[ngulum. . . .quo *illeg.* D
8 acutum *tr. B post* angulum / subten [dens . . . . acutum *illeg. D*
8–9 *post* continentibus *scr. B* lateribus
9 trigo[num . . . . datam *illeg. D*
10–11 et. . . .BG *illeg. D*
12 ab[2]. . . .GB *illeg. D* / et *om. B* / *post* est *scr. B* etiam / GB:BG *B*
14 est[1]: etiam *B* / ABD. . . .et[2] *illeg. D*
15 est[1]: etiam *B* / est[2]: etiam *B*
15–16 Proportio. . . .eius *illeg. D*
17 Et. . . .Sed *illeg. D* / ergo *om. B* / sub[1] *om. B* / *post* Sed *scr. B* et / *post* quod[2] *scr. B* est
18 bis *om. B*
18–19 sub. . . .propor]tio *illeg. D* / GB: BG *B*
19 GB:BG *B*
20 AG. . . .Quo *illeg. D* / *post* ab[2] *bis B* AG quadratum eis que / ergo *om. B*
21–22 ad. . . .datam *illeg. D*

And $AD:DB = (AD \times BG):(DB \times BG)$. Hence the ratio $(DA \times BG)$: $(DB \times BG)$ is also given. Therefore the ratio $2(DB \times BG):(AD \times BG)$ is given. But the ratio $(DA \times BG): \Delta ABG$ is given. Therefore the ratio $2(DB \times BG):\Delta ABG$ is also given. And $2(DB \times BG)$ is that space by which the square of *AG* is greater than the squares of *AB* and *BG*. Therefore that space will have a given ratio to the triangle *ABG*.

LXV. If a triangle has a given acute angle, that space by which the square of the side subtending the acute angle is less than the square of the sides containing the same acute angle has a given ratio to the triangle.

Let the acute-angle triangle *ABG* have a given angle *ABG*. Let the perpendicular *AD* be drawn from the point *A* to *BG*. I say that the space by which the square of *AG* is less than the squares of *AB* and *BG*, that is, the double the rectangle contained by *GB* and *BD*, has a given ratio to the triangle *ABG*.

Since the angle *ABD* is given and the angle *ADB* is also given, hence the remaining angle *BAD* is also given. Therefore the triagnle *ABD* is given in species. Therefore the ratio $BD:DA$ is given. Hence the ratio $(GB \times BD)$: $(GB \times AD)$ is also given. Therefore the ratio $2(GB \times BD):(GB \times AD)$ is given. But the ratio $(GB \times AD): \Delta ABG$ is given. Therefore $2(GB \times BD)$: ΔABG is also given. And $2(GB \times BD)$ is that by which the square of *AG* is less than the squares of *AB* and *BG*. Therefore that space by which the square of *AG* is less than the squares of *AB* and *BG* has a given ratio to the triangle *ABG*.

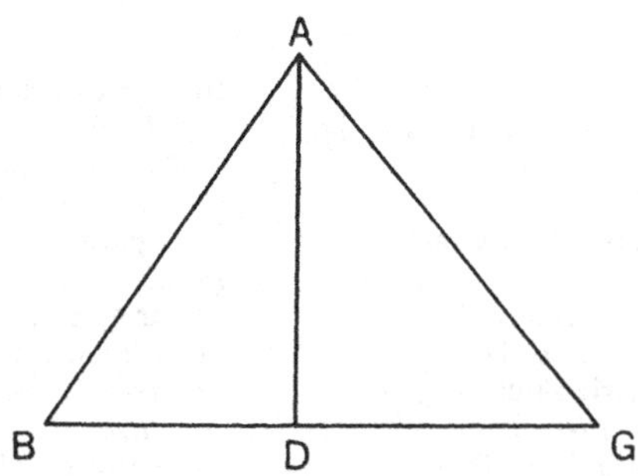

LXVI. Si trigonus datum habeat angulum, quod sub datum angulum continentibus rectis orthogonium ad trigonum proportionem habet datam.

Esto trigonus *ABG* datum habens angulum ad punctum *A*. Dico quo-
(67r) niam quod sub/*BAG* continetur rectangulum ad *ABG* trigonum proportionem habet datam.

Trahatur enim ab *B* puncto super *AG* cathetus *BD*. Quoniam ergo datus est *BAG* angulus, est autem et *ADB* angulus datus, et reliquus ergo *ABD* angulus datus est. Datus est ergo *ABD* trigonus specie. Proportio ergo *AB* ad *BD* data. Sed sicut *AB* ad *BD* ita quod sub *BA* et *AG* ad illud quod sub *BD* et *AG*. Quare et eius quod sub *BAG* ad illud quod sub *BD* et *AG* proportio est data. At vero eius quod sub *AG* et *BD* ad *ABG* trigonum proportio est data. Et eius ergo quod sub *BAG* ad *ABG* trigonum proportio est data.

LXVII. Si trigonus datum habeat angulum, quo maius possunt datum angulum continentia latera ut unum eo quod sub reliquo, illud spatium ad trigonum proportionem habet datam.

Esto trigonum *ABG* datum habens angulum *BAG*. Dico quoniam, quo maius est quod a coutroque *BAG* eo quod ab *BG*, illud spatium ad *ABG* trigonum proportionem habet datam.

Protrahatur enim in directo recte *AB* recta *AD* et ponatur *AG* recte equalis recta *AD* et coniuncta *DG* protrahatur super *E* et trahatur per *E* recte *AG* parallelos *BE*. Et quoniam equalis *AD* recte *AG*, equalis est ergo

1 LXVI:65[a] *B*
1–2 sub . . . . continentibus *illeg. D*
2 orthogonium: ortogonium *D* hortogonium *P*
2–4 proport[ionem . . . . angulum *illeg. D*
5–6 continetur. . . .da]tam *illeg. D*
7–8 sup[er. . . .est[2] *illeg. D*
8–9 reliq[uus. . . .trigo]nus *illeg. D*
10–11 ad[1]. . . .sub[1] *illeg. D*
11–12 quod[1]. . . .sub *illeg. D*
12–13 trigonum. . . .trigonum *illeg. D*
15 LXVII: 66[a] *B* / habeat: habet *O* / possunt: possint *P*
15–17 datum . . . . spatium *illeg. D*
16 continentia *correxi ex* continencia *BOP*
17–19 trigo[num . . . . ab *illeg. D*
19 coutroque: coutraque *O* / ab:a *B* / spatium *om. B* / ad: autem *B*
19–22 spatium. . . .re]cta *illeg. D*
20 trigonum *om. B*
21 directo: directe *B* / AD:AG *B*
22 recta: recte *B* / et[1]. . . .trahatur *illeg. D* / *post* et[1] *bis P* ponatur AG recte equalis recta AD et
23 *post* AG[2] *add. mg. manu rec. O* *4 litteras illeg.*
23–1(p.s) rec[te[1]. . . .protra]cta *illeg. D*

LXVI. If a triangle has a given angle, the rectangle composed of the lines containing the given angle has a given ratio to the triangle.

Let the triangle ABG have a given angle at the point A. I say that the rectangle which is contained by BA and AG has a given ratio to the triangle ABG.

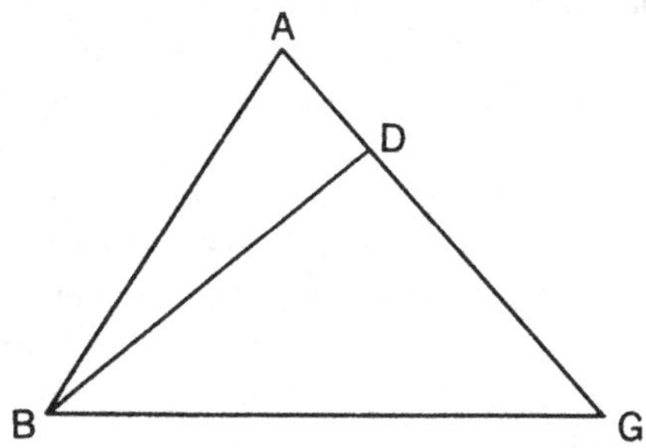

Let the perpendicular BD be drawn from the point B to AG. Since the angle BAG is given and the angle ADB is given, therefore the remaining angle ABD is given. Therefore the triangle ABD is given in species. Therefore the ratio $AB{:}BD$ is given. But $AB{:}BD = (BA \times AG){:}(BD \times AG)$. Hence the ratio $(BA \times AG){:}(BD \times AG)$ is given. But the ratio $(AG \times BD){:}\Delta ABG$ is given. Therefore the ratio of the rectangle contained by BA and AG to the triangle ABG is given.

LXVII. If a triangle has a given angle, that space by which the square of the lines composed of the sides that contain the given angle is greater than the square of the remaining side has a given ratio to the triangle.

Let the triangle ABG have a given angle BAG. I say that the space by which the square of the lines composed of BA and AG is greater than the square of BG has a given ratio to the triangle ABG.

Let AD be the direct extension of AB and let AD be made equal to AG;

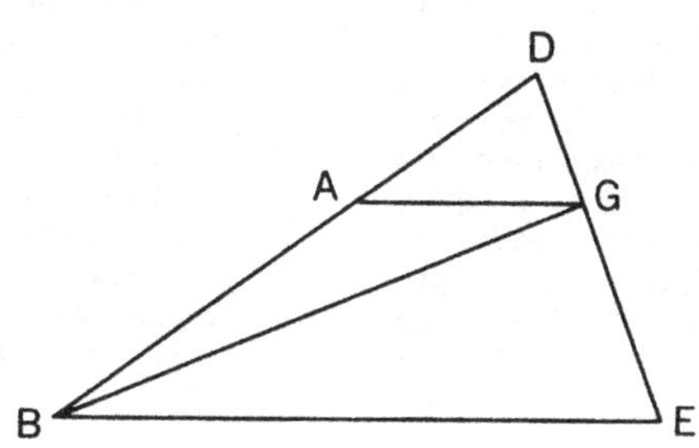

et *DB* recte *BE*. Et protracta est quedam recta *BG*. Quod ergo sub his *DGE* cum eo quod ab *BG* equale est ei quod ab *BD*. Equalis autem recta *AD* recte *AG*. Quod ergo a coutraque *BA* et *AG* equale est ei quod sub *DG* et *GE* cum eo quod ab *BG*. Quare quod a coutraque *BAG* eo quod ab *BG* maius est eo quod sub *DGE*. Dico vero quoniam quod sub his scilicet *DGE* ad *ABG* trigonum proportio est data. Quoniam enim datus est *BAG* angulus et qui deinceps ergo *DAG* datus est. Est autem et uterque istorum scilicet *ADG*, *DGA* datus, medietates enim sunt anguli *BAG*. Datus est enim *BAG*. Datus est ergo *DAG* trigonus specie. Proportio ergo *DA* ad *DG* data. Quare et eius quod ab *AD* ad illud quod ab *DG* proportio est data. Et quoniam sicut *BA* ad *AD* ita *EG* ad *GD*, sed sicut quidem *BA* ad *AD* ita quod sub *BA* et *AD* ad illud quod ab *AD*, sicut vero *EG* ad *GD* ita quod sub *EG*, *GD* ad illud quod ab *GD*, et sicut ergo quod sub *BA*, *AD* ad illud quod ab *DA* ita quod sub *EG*, *GD* ad illud quod ab *GD*. Et permutatim ergo <sicut> quod sub *BA*, *AD* ad illud quod sub *EG*, *GD* ita quod ab *AD* ad illud quod ab *DG*. Proportio autem eius quod ab *AD* ad illud quod ab *DG* data. Proportio ergo et eius quod sub *BAG* ad illud quod sub *EGD* data. Equalis autem est *DA* recte *AG*. Proportio ergo et eius quod sub *BAG* ad illud quod sub *EGD* data. Eius vero quod sub *BAG* ad *ABG* trigonum proportio est data, propter datum esse *BAG* angulum. Et eius ergo quod sub *DGE* ad *ABG* proportio est data. Et est quod sub *DG*, *GE* quo maius est quod a coutraque *BAG* eo quod ab *BG*. Quo ergo maius est quod a coutraque *BAG* eo quod ab *BG*, illud spatium ad trigonum proportionem habet datam.

1 *de* et *add. mg. O* per IIII VI[ti] elementorum / *de* Quod *add. mg. infer. B* Hoce sic ostenditur: Dimittatur a puncto B perpendicularis super DE, cadatque in punctum Z. Patet quod Z est medius punctus DE linee. Quadratum ergo BD valet quadrata DZ et ZB; sed quadratum DZ valet quadrata DG et GZ et id quod fit ex DG in GZ bis. Quadratum vero BG valet quadrata GZ et BZ. Quadratum ergo BD valet quadratum DG et id quod fit ex DG in GZ bis et quadratum BG. Sumatur autem a Z puncto versus E linea equalis ZG. sitque ZH. Relinquitur HE, AD, GD equalis. Quadratum ergo GD est equale ei quod fit ex GD in HE et quod fit ex GD in GZ bis est equale ei quod fit ex GD in HE. Et quod fit ex GD in GZ bis est equale ei quod fit ex GD in GH. Ergo quadratum GD et quod fit ex GD in GZ bis sunt equalia ei quod fit ex GD in totam GE. Quadratum ergo BD valet quadratum BG et id quod fit ex GD in GE.

1–2 quedam. . . .autem *illeg. D*

2 DGE: DG, GE *B* / *de* ei quod *add. mg. O* qui et dulcarnon et V secundi elementorum et quem dulcarnon argumentare, ducta ab puncto super DE perpendiculari

then with DG joined, let it be extended to E and through E let BE be drawn parallel to the straight line AG. Since $AD = AG$, therefore $DB = BE$. And some straight line BG has been drawn. Therefore, $(DG \times GE) + BG^2 = BD^2$. But $AD = AG$. Therefore $(BA + AG)^2 = (DG \times GE) + BG^2$. Hence $(BA + AG)^2$ is greater than BG^2 by $(DG \times GE)$. I say that the ratio $(DG \times GE):\Delta ABG$ is given. Since the angle BAG is given, the adjacent angle DAG is also given. But each of the angles ADG and DGA is given, for they are half of the angle BAG. For the angle BAG is given. Therefore ΔDAG is given in species. Therefore the ratio $DA:DG$ is given. Hence the ratio $AD^2:DG^2$ is also given. Since $BA:AD = EG:GD$ and $BA:AD = (BA \times AD):AD^2$ and also $EG:GD = (EG \times GD):GD^2$, therefore $(BA \times AD):DA^2 = (EG \times GD):GD^2$. By permutation, therefore, $(BA \times AD):(EG \times GD) = AD^2:DG^2$. But the ratio $AD^2:DG^2$ is given. Therefore the ratio $(BA \times AD):(EG \times GD)$ is given. Now $DA = AG$. Therefore the ratio $(BA \times AG):(EG \times GD)$ is given. But the ratio $(BA \times AG):\Delta ABG$ is given, because the angle BAG is given. Therefore the ratio $(DG \times GE):\Delta ABG$ is also given. And $(DG \times GE)$ is the space by which the square of the lines composed of BA and AG is greater than the square of BG. Therefore the space by which $(BA + AG)^2$ is greater than BG^2 has a given ratio to the triangle.

2 recta: reliqua *B*
3 recte: relique *B* / coutraque: coutroque *BP*
3–4 AD....ab[1] *illeg. D*
4 coutraque: coutroque *O*
4–5 Quare....his *illeg. D*
5 DGE:GD, DE *B* / quod *om. B*
6 DGE: DG, GE *B*
6–7 D[GE....DAG *illeg. D*
7 est *om. B* / Est *om. B*
7–8 d[atus....en]im *illeg. D*
8 est *om. P*
8–9 Datus....BAG *illeg. D*
9 est *om. P*
8–10 BAG....data[2] *illeg. D*
10 *de* eius *add. mg. manu rec. O* per XXXVI II libri euclidis / AD:AB *B* / *post* quod[2] *scr. B* est
11 BA[1]:AB *B* / AD[2]:DA *O*
11–12 quidem....quod[2] *illeg. D*
13 ergo *om. B* / ad[2] *bis B*
13–14 GD[1]....EG *illeg. D*
14–15 illud....EG *illeg. D*
16 ab[1]: su *B* / Proportio autem: data est autem proportio *B*
16–17 il[lud[1]....illud *illeg. D*
17 data *om. B*
17–18 Proportio....data *om. B*
18 est *om. B*
18–19 recte....sub[1] *illeg. D*
19 EGD:EG *B* / *post* ad[2] *add. mg. O* per premissam
19–20 trigonum....ergo *illeg. D*
20 propter: proportio *B*
21–22 data....coutraque *illeg. D*
22 coutraque: coutroque *BP* / ergo *tr. B post* est[2]
22–23 est[2]....illud *illeg. D*
23 coutraque: coutroque *BP* / ab BG: ABG *B*
23–24 ad....datam *om. B*

Aliter

Ordinentur enim eadem eis que prius et trahatur a puncto *A* cathetus *AZ* et coniungantur *A D*. Et quoniam datus est *BAG* angulus, et est eius medietas *AGZ*. Est autem et *AZG* datus. Datus est ergo *AZG* trigonus specie. Proportio ergo *AZ* ad *ZG* data. At vero *ZG* ad *GE* proportio est data, dupla enim est eius. Et *EG* ergo ad *AZ* proportio est data. Quare et eius quod sub *EG*, *GD* ad illud quod sub *AZ*, *GD*, proportio est data. Eius vero sub *AZ*, *GD* ad *AGD* trigonum proportio est data, duplum enim est eius. Et eius ergo quod sub *EG*, *GD* ad *AGD* trigonum proportio est data. Equalis autem *AGD* trigonus *AGB* trigono, super eandem enim basem sunt que est *AG* et in eisdem parallelis que sunt *AG* et *BD*. Et eius ergo quod sub *EG*, *GD* ad *ABG* trigonum proportio est data. Et est quod sub *EG*, *GD* quo maius est quod a coutraque *BAG* eo quod ab *BG*. Quo ergo maius est quod a coutraque *BA*, *AG* eo quod ab *GB*, illud spatium ad trigonum proportionem habet datam.

Aliter

A enim angulus vel rectus est vel actus vel obtusus. Esto primum rectus. Quod ergo a coutraque *BAG* eo quod ab *BG* superhabundat ei quod bis sub *BA*, *AG*. Et est eius quod bis sub *BA*, *AG* ad *ABG* trigonum proportio data.

Esto vero acutus *BAG* et trahatur a puncto *G* super *AB* cathetus *GD*. Quoniam ergo oxigonius est *ABG* trigonus et cathetus tracta est *GD*, que ergo

2 Ordinentur: Ordinetur *B* / Ordinentur....trahatur *illeg. D* / *post* puncto *scr. B* ad trigonum proportion aliter datam
3 est² *bis P*
3–4 da[tus....medietas *illeg. D*
4 AZG¹: AGZ *P* / est *tr. B post* ergo
5 specie....vero *illeg. D*
6 data *tr. B ante* proportio *ln. 5*
6–7 ad....eius *illeg. D*
7 EG:GE *B* / post GD² *del. et scr. D* ad AGD trigonum
8 vero....GD *om. D* / ad AGD *bis B* / duplum: dupla *B*
8–9 duplum....ergo *illeg. D*
9 ad AGD *bis B* / post data *bis D* [duplum enim est eius. Et eius ergo] quod sub EG GD ad AGD trigonum proportio est data.
10 autem: enim *B* / basem: basim *OP*
10–11 trigo[no....est *illeg. D*
11 in *om. P*
12 GD....data *illeg. D* / sub *om. B*
13 coutraque: coutroque *B* / ergo *om. B*
13–14 eo.... quod² *illeg. D*
14–17 proporti[onem....angu]lus *illeg. D*
18 Quod....superhabun]dat *illeg. D* / superhabundat: superabundat *B hic et ubique*
19 *post* sub¹ *scr. P* his
19–20 ad....data *illeg. D*
21–22 GD....ABG *illeg. D*
22 est¹ *om. B*

Another proof

Let the same arrangement hold as in the first demonstration. Let there be drawn from the point A [to EG] a perpendicular AZ and let A and D be joined.

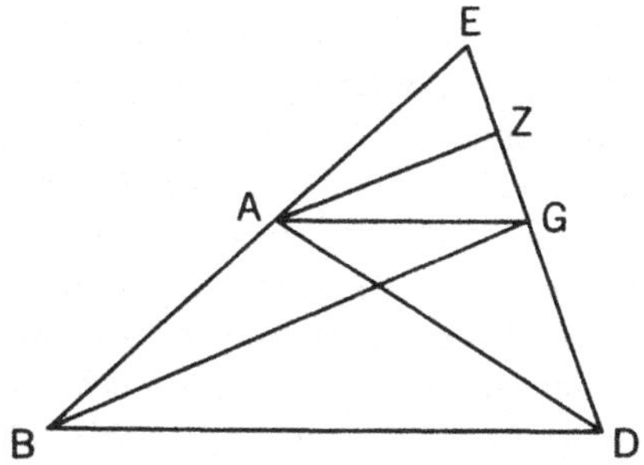

Since the angle BAG is given, its half, the angle AGZ, is also given. Now the angle AZG is given. Therefore the triangle AZG is given in species. Therefore the ratio $AZ{:}ZG$ is given. But the ratio $ZG{:}GE$ is given, for $GE = 2ZG$. Therefore the ratio $EG{:}AZ$ is given. Hence the ratio $(EG \times GD){:}(AZ \times GD)$ is given. But the ratio $(AZ \times GD){:}\Delta AGD$ is given, for $(AZ \times GD) = 2\Delta AGD$. Therefore the ratio $(EG \times GD){:}\Delta AGD$ is given. But $\Delta AGD = \Delta ABG$, for they are on the same basis AG and between the same parallel lines AG and BD. Therefore the ratio $(EG \times GD){:}\Delta ABG$ is given. And $(EG \times GD)$ is the space by which $(BA + AG)^2$ is greater than BG^2. Therefore that space by which $(BA + AG)^2$ is greater than BG^2, has a given ratio to the triangle.

Another proof

The [given] angle A is either a right, acute or obtuse angle. First, let it be a right angle. Then $(BA + AG)^2$ is greater than BG^2 by $2(BA \times AG)$. And the ratio $2(BA \times AG){:}\Delta ABG$ is given.

Now let the angle BAG be acute and from the point G let the perpendicular GD be drawn to AB. Since the triangle ABG is an acute angle and the perpendicular GD has been drawn, $BA^2 + AG^2 = BG^2 + 2(BA \times AD)$.

ab *BAG* equalia sunt ei quod ab *BG* et ei quod bis sub *BA*, *AD*. Commune apponatur quod bis sub duabus *BAG*. Que ergo ab his scilicet *BAG* cum eo quod bis sub duabus *BAG*, quod est illud quod a coutraque *BAG*, equalia sunt et ei quod ab *BG* et ei quod bis sub duabus *BAD* et adhuc ei quod bis sub duabus *BAG*, hoc est, ei quod bis sub coutraque *GAD* et *AB*. Quare (67v) quod a coutraque *BAG* maius est eo quod ab *BG* eo/quod bis sub coutraque *DAG* et *BA*. Et quoniam datus est *BAG* angulus, est autem et *ADG* angulus datus, et reliquus ergo *DGA* est datus. Datus est ergo *ADG* trigonus specie. Proportio ergo *AD* ad *AG* data. Quare et coutriusque *DAG* ad *AG* proportio est data. Et eius ergo quod sub coutraque *DAG* et *AB* ad illud quod sub his scilicet *BAG* proportio est data. Et eius quod bis sub coutraque *DAG* et *AB* ad illud quod sub *BAG* proportio est data. Atqui eius quod sub his scilicet *BAG* ad *BAG* trigonum proportio est data, per datum esse ad *A* angulum. Et eius ergo quod bis sub coutraque *DAG* et *AB* ad *ABG* trigonum proportio est data.

Sed obtusus esto *BAG* et eiecta *BA* trahatur super eadem cathetus *GE* et ponatur *AE* equalis recta *AZ*. Quoniam ergo obtusus est *BAG* angulus et

1 *post* sub *add. mg. O* per XIII secundi elementorum / BA, AD: BAD *B* / quod[1]....Commune *illeg. D*
2 his: hiis *D*
2–3 cum....est *illeg. D*
3 *post* quod[3] *scr. B* est
4 quod[1]: que *D* / ab....quod[2] *om. B*
4–5 duabus....dua]bus *illeg. D*
5 *post* ei quod *add. mg. O* per I secundi elementorum / ei *om. B*
6 co[utraque....quod[2] *illeg. D* / ab ....quod *om. B*
7–8 autem....DGA *illeg. D*
8 DGA: DAG *B* / Datus est: Est datus *B*
9 AD....DAG *illeg. D* / coutriusque: coutraque *B*
10 ergo *om. B* / *post* et *add. mg. O* per I VI[ti] elementorum
11 ad....data *lleg. D*
12 BAG:AG *B*
12–13 sub....ad *illeg. D*
13 his: ms(?) *B* / *post* BAG[2] *add. mg. O* per premissam
14–15 esse....AB *illeg. D*
15 AB ad *om. B*
16 eadem: eam *B*
16–17 esto....pona]tur *illeg. D* / et ....BAG *om. P*
17 AZ:AG *B* / est *om. B*
17–1(p.s.) obtusus....cum *illeg. D*

Let $2(BA \times AG)$ be added commonly. Therefore:

$$BA^2 + AG^2 + 2(BA \times AG) = (BA + AG)^2$$
$$= BG^2 + 2(BA \times AD) + 2\,(BA \times AG)$$
$$= BG^2 + 2(GA + AD) \times AB$$

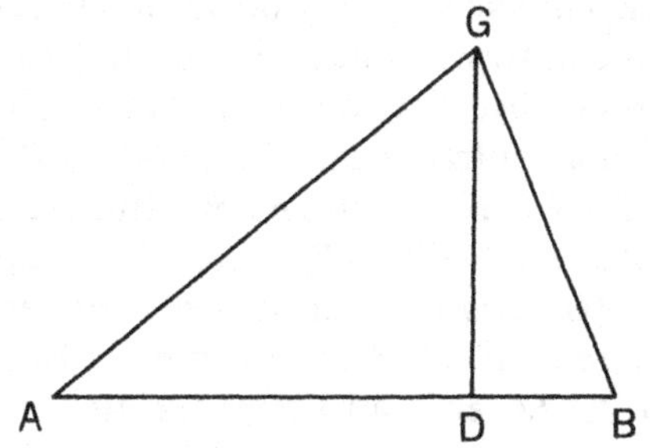

Therefore $(BA + AG)^2$ is greater than BG^2 by $2(DA + AG) \times BA$. And since the angle BAG is given and the angle ADG is given, hence the remaining angle DGA is also given. Therefore the triangle ADG is given in species. Therefore the ratio $AD{:}AG$ is given. Hence the ratio $(DA + AG){:}\ AG$ is also given. Therefore the ratio $(DA + AG) \times AB{:}(BA \times AG)$ is given. And the ratio $2(DA + AG) \times AB{:}(BA \times AG)$ is given. Now $(BA \times AG){:}\Delta BAG$ is given, because the angle at A is given. Therefore the ratio $2(DA + AG) \times AB{:}\Delta ABG$ is also given.

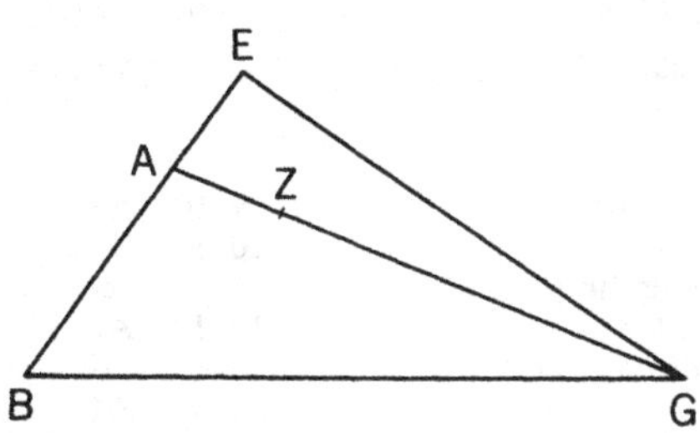

But let the angle BAG be obtuse, and with BA extended let the perpendicular GE be drawn to the same [extension of BA] and let AZ be equal to AE. Since the angle BAG is obtuse and the perpendicular GE has been drawn, therefore

$$BA^2 + AG^2 + 2(BA \times AE) = BA^2 + AG^2 + 2(BA \times AZ)$$
$$= BG^2$$

Let $2(BA \times AG)$ be added commonly.

cathetus tracta est *GE*, que ergo ab *BA* et *AG* cum eo quod bis sub his scilicet *BAE*, hoc est, quod bis sub duabus *BAZ*, equale est ei quod ab *BG*. Commune apponatur quod bis sub duabus *BAG*. Que ergo ab *BA*, *AG* cum eo quod bis sub duabus *BAG*, hoc est, quod a coutraque *BAG* cum eo quod bis sub duabus *BAZ*, equalia sunt ei quod ab *BG* cum eo quod bis sub duabus *BAG*. Commune auferatur quod bis sub duabus *BAZ*. Quod ergo a coutraque *BAG* equale est ei quod ab *BG* et ei quod bis sub duabus *BA GZ*. Quare quod a coutraque *BAG* ei quod ab *BG* superhabundat eo quod sub duabus *BA GZ* bis. Et quoniam datus est *BAG* angulus et angulus ergo *EAG* datus est. Sed et *GEA* datus. Et reliquus ergo *AGE* datus est. Datus est ergo *AGE* trigonus specie. Proportio ergo *GA* ad *AE* data. hoc est, ad *AZ*. Quare et *AG* ad *ZG* proportio est data. At vero *AG* ad *GE* proportio est data. Et *EG* ergo ad *GZ* proportio est data. Quare et eius quod sub *EG*, *AB* ad illud quod sub *GZ*, *AB* proportio est data. Eius quoque quod sub *AB*, *GE* ad *ABG* trigonum proportio est data. Quare et eius quod bis sub duabus *GZ*, *BA* ad *ABG* trigonum proportio est data. Et est quod bis sub duabus *GZ*, *BA* quo maius est quod a coutraque *BAG* eo quod ab *BG*. Quo ergo quod a coutraque *BAG* maius est eo quod ab *BG* illud spatium ad trigonum proportionem habet datam.

Aliter

Protrahatur *BA* super *D* et sumatur recte *GA* equalis *AD* et coniungantur *D*, *G*. Quoniam ergo datus est *BAG* angulus et est eius medietas uterque horum *ADG* et *AGD*, datus est ergo uterque *ADG* et *AGD*. Et reliquus ergo *DAG* datus est. Datus est ergo *AGD* trigonus specie. Proportio ergo

1 que: quod *B* / his: hiis *D*
2 quod[1] *om. B*
2–3 BAE....apponatur *illeg. D*
3 BA:GA *B* / AG:A *B*
3–5 Que....quod[2] *illeg. D*
4 duabus *om. B*
4–5 BAG[1]....duabus *om. P et om.(?) D*
5–6 cum....BAG *add. mg. O et om. BP*
5–7 duabus[2]....cou]traque *illeg. D*
6 auferatur: auferantur *B*
7 bis *om. P*
7–8 quod[1]....BG *illeg. D*
8 BA GZ:BAZ *B* / *post* GZ *scr. et del. O* bis et quoniam datus. / BAG: DAG *P*
9–10 sub....est *illeg. D*
10–11 Et....AE *illeg. D*
11 est[1] *om. B*
12–13 AZ....ad *illeg. D*
13–15 propor[tio[2]....quoque *illeg. D*
15–16 AB....ABG *illeg. D*
16 bis *om. B* / ABG:AB *B*
16–18 trigo[num....quod[1] *illeg. D*
17 maius *corr. B ex* magnum / *post* BAG *bis P* maius
18–19 Quo....datam *illeg. D*
19 habet *om. B*
20–22 Aliter....datus *illeg. D*
21–22 coniungantur: coniungan *B*
22–24 angulus....ergo[1] *illeg. D*
24–1(p.s.) datus....est *illeg. D*

Then $BA^2 + AG^2 + 2(BA \times AZ) + 2(BA \times AG) = (BA + AG)^2 + 2(BA \times AZ) = BG^2 + 2(BA \times AG)$

Let $2(BA \times AZ)$ be subtracted commonly.

Therefore $(BA + AG)^2 = BG^2 + 2\,(BA \times GZ)$

Therefore $(BA + AZ)^2$ exceeds BG^2 by $2(BA \times GZ)$.

Since the angle *BAG* is given, the angle *EAG* is also given. But the angle *GEA* is also given. Therefore the remaining angle *AGE* is given. Therefore the triangle *AGE* is given in species. Therefore the ratio *GA*:*AE*, that is, *GA*:*AZ* is given. Hence the ratio *AG*:*ZG* is also given. But the ratio *AG*: *GE* is given. Therefore the ratio *EG*:*GZ* is given. Hence the ratio $(EG \times AB):(GZ \times AB)$ is also given. But the ratio $(AB \times GE):\Delta ABG$ is given. Hence the ratio $2(GZ \times BA):\Delta ABG$ is also given. And $2(GZ \times BA)$ is that space by which $(BA + AG)^2$ is greater than BG^2. Therefore that space by which $(BA + AG)^2$ is greater than BG^2 has a given ratio to the triangle.

Another proof

Let *BA* be protracted to *D* and let *AD* be taken equal to *GA* and let *D* and *G* be joined. Since the angle *BAG* is given and each of the angles

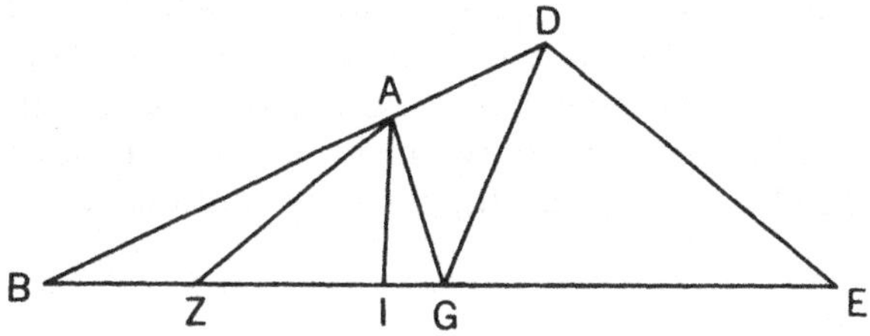

ADG and *AGD* is half of it, therefore each angle of *ADG* and *AGD* is given. Therefore the remaining angle *DAG* is also given. Therefore the triangle *AGD* is given in species. Therefore the ratio *AG*:*GD* is given. Since the angle *ADG* is given, let each angle of *DEG* and *AZG* be put equal to the angle *ADG*. Since $\angle BDG = \angle DEG$ and the angle *ABG* is common to the triangles *DBE* and *DBG*, the remaining angle *BDE* is equal to the remaining angle *BGD*. Therefore the triangle *BDE* is equiangular with the triangle *DBG*. Therefore $EB{:}BD = BD{:}BG$. Therefore $EB \times BG = BD^2$, that is,

$$(EG \times GB) + GB^2 = BD^2$$
$$= (BA + AG)^2, \quad \text{for } DA = AG.$$

AG ad *GD* data. Et quoniam datus est *ADG*, subiaceat ei equalis uterque *DEG* et *AZG*. Et quoniam equalis est *BDG* ei qui est *DEG*, communis vero *ABG* trigoni *DBE* existens et *DBG*, reliquus ergo *BDE* reliquo *BGD* est equalis. Equiangulus ergo est *BDE* trigonus *DBG* trigono. Est ergo sicut *EB* ad *BD* ita *BD* ad *BG*. Quod ergo sub *EB* et *BG*, hoc est, sub *EGB* cum eo quod ab *GB*, equale est ei quod ab *BD*, hoc est, ei quod a coutraque *BAG*, equalis est enim *DA* recta recte *AG*. Quod ergo sub *EGB* cum eo quod ab *GB* equale est ei quod a coutraque *BAG*. Quod ergo a coutraque *BAG* eo quod ab *BG* superhabundat ei quod sub *BG*, *GE*. Dico ergo quod proportio est eius quod sub *BG*, *GE* ad *ABG* trigonum data. Quoniam enim equalis est *BDE* angulus *BGD*, quorum *ADG* angulus angulo *AGD* equalis, reliquus ergo *GDE* reliquo *AGB* equalis est. Est autem et *DEG* angulus angulo *AZG* equalis. Reliquus ergo *GAZ* reliquo *DGE* est equalis. Equiangulus ergo *AZG* trigonus *DEG* trigono. Est ergo sicut *GA* ad *AZ* ita *DG* ad *GE*. Et permutatim ergo sicut *AG* ad *GD* ita *AZ* ad *GE* proportio autem *AG* ad *GD* data. Proportio ergo et *AZ* ad *GE* data. Trahatur ab *A* super *BG* cathetus *AI*. Et quoniam datus est *AZG*, est autem et *AIZ* datus, reliquus ergo *ZAI* datus est. Datus est ergo *AIZ* trigonus specie. Proportio ergo *ZA* ad *AI* data. At vero *ZA* ad *GE* proportio est data. Et *AI* ergo ad *GE* proportio est data. Quare et eius quod sub *AI*, *BG* ad illud quod sub *BG*, *GE* proportio est data. Eius quoque quod sub *AI*, *BG* ad *ABG* trigonum proportio est data. Et eius quod sub *BG*, *GE* ergo ad *ABG* trigonum proportio est data. Et est quod sub *BG*, *GE* quo maius est quod a coutraque *BAG* eo quod ab *BG*. Quo ergo maius est quod a coutraque *BAG* eo quod ab *BG*, illud spatium ad trigonum proportionem habet datam.

1 *supra* ADG *add. O* angulus / *de* subiaceat *add. mg. infer. O* (*cf. commentum in B p.166*) Hoc sic ostenditur: Demittatur a puncto B perpendicularis DE et cadat ad punctum Z. Patet quod Z est medius punctus DE linee. Quadratum autem BD valet quadrata DZ et ZB; sed quadratum DZ valet quadrata DG et GZ et id quod fit ex DG in GZ bis. Quadratum vero BG valet quadrata GZ et BZ. Quadratum ergo BD valet quadratum DG et id quod fit ex ductu DG in GZ bis et quadratum BG. Sumatur autem a Z puncto versus E linea equalis ZG, sitque ZH. Relinquitur HE, AD, GD equalis. Quadratum GD est equale ei quod fit ex GD in HE et quod fit ex GD in GZ bis est equale ei quod fit ex GD in GH. Ergo quadratum DG et quod fit ex GD in GZ bis sunt equalia ei quod fit ex GD in totam GE. Quadratum ergo BD valet quadratum BG et id quod fit ex GD in GE.

1–3 subiaceat. . . . trigoni *illeg. D*

2 est[1] *om. B* / qui: quod *B*

3 DBG: DBD *B* / est *tr. B post* equalis *ln. 4*

3–4 existens. . . . tri]gonus *illeg. D*

Therefore, $(EG \times GB) + GB^2 = (BA + AG)^2$.
Therefore, $(BA + AG)^2$ exceeds BG^2 by $BG \times GE$.

I say, therefore, that $BG \times GE$ has a given ratio to the triangle ABG. Since $\angle BDE = \angle BGD$ and its part $\angle ADG = \angle AGD$, hence the remaining angle GDE is equal to the remaining angle AGB. But also $\angle DEG = \angle AZG$. Therefore the remaining angle GAZ is equal to the remaining angle DGE. Therefore the triangle AZG is equiangular with the triangle DEG. Therefore $GA:AZ = DG:GE$. Therefore, by permutation, $AG:GD = AZ:GE$. Now the ratio $AG:GD$ is given. Therefore the ratio $AZ:GE$ is given. Let the perpendicular AI be drawn from A to BG. Since the angle AZG is given and the angle AIZ is also given, therefore the remaining angle ZAI is given. Therefore the triangle AIZ is given in species. Therefore the ratio $ZA:AI$ is given. But the ratio $ZA:GE$ is given. Therefore the ratio $AI:GE$ is given. Hence the ratio $(AI \times BG):(BG \times GE)$ is given. The ratio $(AI \times BG):\Delta ABG$ is also given. Therefore the ratio $(BG \times GE):\Delta ABG$ is given. And the rectangle $BG \times GE$ is the space by which $(BA + AG)^2$ greater than BG^2. Therefore that space by which $(BA + AG)^2$ is greater than BG^2 has a given ratio to the triangle

4 ergo[1] *om. B*
4–6 DBG....ab[1] *illeg. D*
5 BD ad BG *lac. BP* / Quod ergo: Ergo quod *B* / EB et BG: EBG *B* / hoc....EGB *om. B*
6 ab[2]....quod *om. D* / est[2] *om. P*
7 DA:DAG *B*
7–8 recta....GB *illeg. D*
8 a[1]:ab *OP*
8–9 cou[traque[2]....GE *illeg. D*
9 ab:ad *B* / GE:TE *B*
10 est *om. B*
10–11 prop[ortio....BGD *illeg. D*
11 BGD:BDG *B* / ADG:AGD *BP*
11–12 angulo....DEG *illeg. D*
13 est *tr. B post* equalis[2]
13–14 Reliqu[us....trigo]nus *illeg. D*
14–15 AZ....ita[2] *illeg. D*
15 *post* AG *scr. B* et
16–17 Proportio....BG *illeg. D*
17 AI *corr. D ex* EI
17–18 et....AIZ *illeg. D*
18 ergo[2]: autem *B*
19 ZA[2]: ZE *B*
19–20 ZA[2]....propor]tio *illeg. D*
20–21 sub[2]....sub *illeg. D*
22–23 GE....data *illeg. D*
24 ergo *om. B*
24–25 eo....datam *illeg. D* / Quo ....illud *om. (?) D*

LXVIII. Si duo equiangula parallelogramma adinvicem proportionem habeant datam, et unum latus ad unum latus proportionem habeat datam, et reliquum latus ad reliquum latus proportionem habebit datam.

(68r) /Duo enim equiangula parallelogramma *AB* et *GD* adinvicem proportionem habeant datam, habeatque latus unum ad unum proportionem datam, sitque *BE* ad *ZD* proportio data. Dico quod et *AE* ad *ZG* proportio est data.

Educatur enim secundum *EB* parallelogrammo *GD* equale parallelogrammum *EI* et iaceat quasi in directo esse *AE* ei que est *ET*. In directo ergo est et *KB* ad *BI*. Quoniam ergo proportio est *AB* ad *GD* data, equale autem est *GD* ei quod est *EI*, proportio ergo et *AB* ad *EI* data. Quare et *AE* ad *ET* porportio est data. Et quoniam equale est *EI* ei quod est *GD*, est vero et equiangulum, eorum ergo que sunt *EI* et *GD* contrarie passa sunt latera circa equales angulos. Est ergo sicut *EB* ad *ZD* ita *ZG* ad *ET*. Proportio autem *EB* ad *ZD* data. Et *ZG* ergo ad *ET* proportio est data. Sed *ET* ad *AE* proportio est data. Et *AE* <ergo> ad *ZG* proportio est data.

Aliter

Adiaceat data recta *K*. Et quoniam proportio est *A* ad *B* data, eadem ei fiat *K* ad *L*. Proportio autem *A* ad *B* data. Proportio ergo *K* ad *L* data. Data autem *K*. Data ergo et *L*. Rursum quoniam proportio est data *GD* ad *EZ*, eadem ei fiat recte *K* ad *M*. Proportio ergo et recte *K* ad *M* data. Data autem *K*. Data ergo et *M*. Est autem et *L* data. Proportio ergo *L* ad

1 LXVIII: 67^{a} *B*
1–2 adin[vicem proportionem *illeg. D*
2 ad unum latus *om. D*
3 et . . . . re]liquum^{2} *illeg. D* / post reliquum^{2} *om. DO* latus
5–6 parallelogramma. . . . pro]portionem^{1} *illeg. D*
6 habeatque. . . . proportionem *illeg. D*
7–9 propor[tio^{2}. . . . EB *illeg. D*
9 EB *correxi ex* EA *BOP*
10 esse: commune *B* / AE ei: AGE *B*
10–11 est. . . . ad^{1} *illeg. D*
11 ergo: igitur *B*
12 autem *tr. B post* est^{1} / EI. . . . data *illeg. D* / EI^{2}:EA *B*
13 *post* EI *scr. P* I / ei *om. B*
14 equiangulum. . . . GD *illeg. D* / *de* EI *add. mg. manu rec. O* per XIIII 6 libri
15 circa: contra *D* / ZD:ZG *B*
15–16 ZD. . . . EB *illeg. D*
16 Et *om. B*
17-18 propor[tio^{1}. . . . est *illeg. D*
20–21 eadem. . . . A *illeg. D*
21 autem *om. B*
22 est data *bis P*
22–23 Rur[sum. . . . EZ *illeg. D*
23 et *om. P*
24 Da[ta^{1}. . . . L^{1} *illeg. D*

LXVIII. If two equiangular parallelograms have a given ratio to each other, and if one side also has a given ratio to one side, the remaining side will have a given ratio to the remaining side.

Let two equiangluar parallelograms *AB* and *GD* have a given ratio to each other, and let one side have a given ratio to one side, that is, let the ratio *BE*:*ZD* be given. I say that the ratio *AE*:*ZG* is also given.

To *EB* let there be applied the parallelogram *EI* equal to the parallelogram *GD*, and let it be posited that *AE* and *TE* are in the same straight line. Therefore *KB* and *BI* are also in the same straight line. Since the ratio *AB*:*GD* is given and $BD = EI$, therefore the ratio *AB*:*EI* is also given. Hence the ratio *AE*:*ET* is also given. And since $EI = GD$ and they are equiangular, hence the sides of *EI* and *GD* about the equal angles are re-

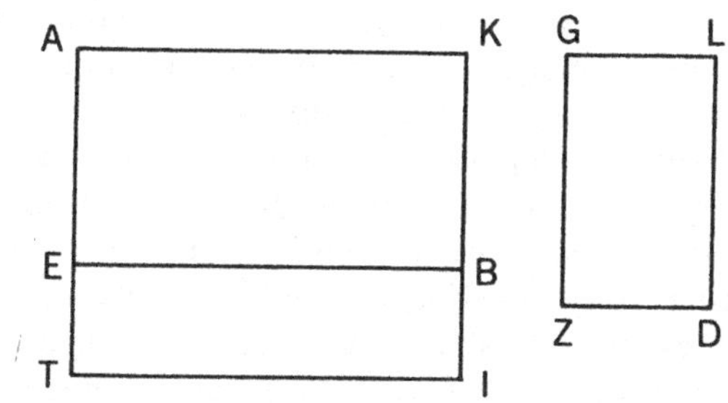

ciprocally proportional. Therefore $EB{:}ZD = ZG{:}ET$. Now the ratio *EB*: *ZD* is given. Therefore the ratio *ZG*:*ET* is also given. But the ratio *ET*: *AE* is given. Therefore the ratio *AE*:*ZG* is also given.

Another proof

Let the straight line *K* be posited. And since the ratio *A*:*B* is given, let the same ratio as it be made of *K* to *L*. Now the ratio *A*:*B* is given. Therefore the ratio *K*:*L* is given. Now *K* is given. Therefore *L* is also given. Further, since the ratio *GD*:*EZ* is given, let the same ratio as it be made of *K* to *M*. Therefore the ratio *K*:*M* is given. Now *K* is given. Therefore *M* is also given. And *L* is also given. Therefore the ratio *L*:*M* is given. And since *A* is equiangular with *B*, *A* has to *B* a compound ratio of the ratios of the sides, that is, of the ratio *GD*:*EZ* and *TG*:*EI*.

[Then, $A{:}B = (GD{:}EZ) \times (TG{:}EI)$]

But also $K{:}L = (K{:}M) \times (M{:}L)$

M data. Et quoniam equiangulum est *A* ei quod est *B*, ergo *A* ad *B* proportionem habet compositam ex proportionibus laterum, hoc est, ex proportione quam habet *GD* ad *EZ* et quam *TG* ad *EI*. Sed et *K* ad *L* proportionem habet compositam ex proportione quam habet *K* ad *M* et quam *M* ad *L*. Composita ergo proportio ex proportionibus quas habet *GD* ad *EZ* et *TG* ad *EI* eadem est composite ex proportionibus quas habet *K* ad *M* et *M* ad *L*, quarum *GD* ad *EZ* proportio eadem est proportioni *K* ad *M*. Reliqua ergo proportio *TG* ad *EI* eadem est proportioni *M* ad *L*. Atqui *M* ad *L* proportio est data. Proportio ergo *TG* ad *EI* data.

LXIX. Si duo parallelogramma datos habeant angulos et proportionem adinvicem habeant datam, latus quoque unum ad unum latus proportionem habeat datam, et reliquum latus ad reliquum proportionem habebit datam.

Duo enim parallelogramma *AB* et *IE* habentia datos angulos et adinvicem proportionem habeant datam, proportio quoque *DB* ad *ZI* sit data. Dico quod et *AD* ad *EZ* proportio data est.

Si ergo equiangulum *AB* parallelogrammum parallelogrammo *EI* manifestum. Si autem non, constituatur ad *DB* et ad eiusdem punctum *D* angulo *EZI* equalis angulus *BDK* et compleatur *DL* parallelogrammum. Et quoniam datus est uterque horum *DAG* et *AKD*, et reliquus ergo *ADK* datus est. Datus est ergo *ADK* trigonus specie. Proportio ergo *AD* ad *DK* data. Et quoniam proportio est *DG* ad *ZT* data, suppositum est enim, et

1–2 est[2]. . . .com]positam *illeg. D*
3 habet. . . .L *illeg. D*
4 ad[1] *om. P*
4–5 hab[et[2]. . .proportio]nibus *illeg. D*
5 L:B *B*
6–7 ead[em. . . .GD *illeg. D*
7 quarum: quare *BP* / proportioni: proportio *B* / K:IK *B*
8 est *om. B* / proportioni: proportio *BP*
8–9 ergo. . . .data[2] *illeg. D*
9 TG *om. P*
10 LXIX: 68[a] *B*
10–11 datos . . . . ad]invicem *illeg. D*
11 unum[2]: aliud *B*
11–12 da[tam . . . . habeat *illeg. D*
12–14 reliqu[um[1] . . . . datos *illeg. D*
14 parallelogramma: paralillos *B*
14–15 adinvicem: invicem *B*
14–16 adinvi[cem. . . .EZ *illeg. D*
15 quoque: ergo *B* / ZI:Z *B*
16 et *om. B* / data[2] *tr. B post* est
17 equiangulum *lac. P* / equiangulum AB *om. B* / EI:I *B*
17–18 ergo. . . .ad[2] *illeg. D*
18 autem:a *B* / DB:BD *B* / eiusdem: eius *B*
18–20 pun[ctum. . . .horum *illeg. D*
20–22 AKD. . . .da]ta[1] *illeg. D*
21 ergo[1] *om. B*
22–1(p.s.) proportio. . . .ZT *illeg. D*

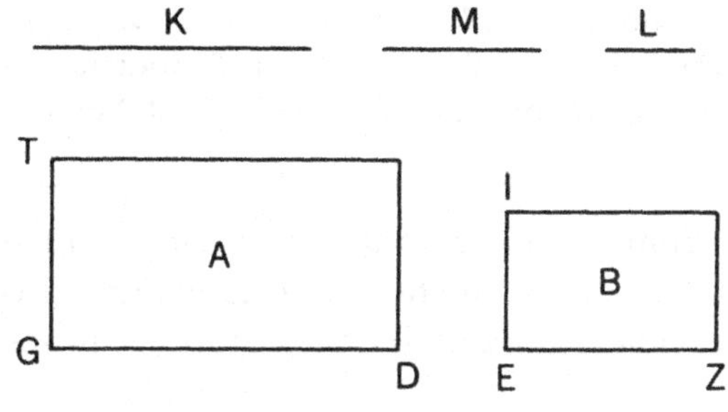

Therefore $(GD{:}EZ) \times (TG{:}EI) = (K{:}M) \times (M{:}L)$, in which $GD{:}EZ = K{:}M$. Therefore the remaining ratio $TG{:}EI$ is equal to the ratio $M{:}L$. Now the ratio $M{:}L$ is given. Therefore the ratio $TG{:}EI$ is given.

LXIX. If two parallelograms have given angles and a given ratio to each other, and if one side also has a given ratio to one side, the remaining side will also have a given ratio to the remaining side.

Let the two parallelograms AB and IE having given angles have also a given ratio to each other, and let the ratio $DB{:}ZI$ be given. I say that the ratio $AD{:}EZ$ is also given.

If the parallelogram AB is equiangular with the parallelogram EI, it is manifest [from the preceeding proposition]. But if AB is not equiangular with EI, on DB and at the same point D let an angle BDK be made equal to the angle EZI and let the parallelogram DL be completed. And since each of the angles DAG and AKD is given, therefore the remaining angle ADK is given. Therefore the triangle ADK is given in species. Therefore the ratio $AD{:}DK$ is given. Since the ratio $DG{:}ZT$ is given, for it has been supposed so, and $DG = DL$, therefore the ratio $DL{:}ZT$ is given. DL is equiangular with ZT and the ratio $DL{:}EI$ is given, and the ratio of the straight lines

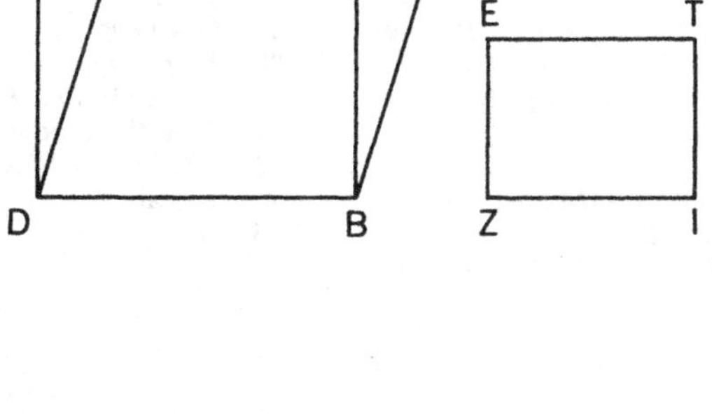

est equale *DG* ei quod est *DL*, proportio ergo *DL* ad *ZT* data. Et equiangulum est *DL* ei quod est *ZT*, et proportio est *DL* ad *EI* data et est *BD* recte ad rectam *ZI*, suppositum est enim. Proportio ergo et *DK* ad *EZ* data. Sed et *DK* ad *AD* proportio est data. Et *AD* ergo ad *EZ* proportio est data.

LXX. Si duorum parallelogrammorum circa equales angulos vel inequales quidem datos autem latera adinvicem proportionem habeant datam, eadem quoque parallelogramma proportionem habebunt datam.

Duorum enim parallelogrammorum *AB* et *EI* circa equales angulos qui ad *Z*, *G* vel circa inequales quidem datos autem latera adinvicem proportionem habeant datam, hoc est, proportio sit *AG* quidem ad *EZ* data, sed et *BG* ad *ZI*. Dico quod et *GD* ad *ZT* proportio est data.

Esto enim equiangulum *GD* ei quod est *ZT*, et educatur secundum *GB* rectam *ZT* parallelogrammo equale rectilineum *GM*, et ponatur quasi in directo esse *AG* ei que est *GN*. Et *DB* ergo ei que est *BM* in directo est. Et est equale *BN* ei quod est *ZT*. Est autem et equiangulum. Istorum ergo videlicet *BN* et *ZT* contrarie patiuntur latera que circa equales angulos. Est ergo sicut *GB* ad *ZI* ita *ZE* ad *GN*. Proportio autem *GB* ad *ZI* data. Proportio ergo et *EZ* ad *GN* data. Sed et *EZ* ad *AG* proportio est data. Et *AG* ergo ad *GN* proportio est data. Quare et *GD* ad *GM* proportio est data. Est autem *GM* ei quod est *ZT* equale. Proportio ergo et *GD* ad *EI* data.

Non sit autem equiangulum *AB* ei quod est *ZT*, et constituatur ad *BG*

1–3 e[quiangulum....ZI *illeg. D*
3–5 suppo[situm....data *illeg. D*
6 LXX: 69[a] *B* / equales *tr. P post* angulos
6–7 duorum.... au]tem *illeg. D*
7–8 habeant: habent *BO*
7–9 latera .... habebunt *illeg. D*
10 Duorum:Duo *B* / parallelogrammorum: paralellogramma *B* paralliilogramma *P*
10–13 Duorum....est *illeg. D*
12 EZ:Z *P*
14 GB:BG *B*
14–15 e[ducatur....GM *illeg. D*
15 quasi: quare *BP*
16 que[1] *et* que[2]: quod *B* / GN: G ei *B* / directo[2]: directum *B*
16–17 Et....est[2] *illeg. D*
17 autem....Istorum *illeg. D*
18 videlicet: scilicet *B* / angulos *om. B*
18–19 et....sicut *illeg. D*
19 ergo *tr. O post* sicut
19–20 autem....A]G *illeg. D* / ad AG *bis P*
20 GN:G *P*
20–21 proportio....AG *om. B* / Et ....data[2] *om. P*
21–22 est[1]....ad *illeg. D*
22 GD:EI B / EI:GD B
23 ad: ab *B*
23–1(p.s.) AB....punc]tum *illeg. D*

BD:*ZI* is given, for it has been supposed so. Then the ratio *DK*:*EZ* is also given. But the ratio *DK*:*AD* is also given. Therefore the ratio *AD*: *EZ* is also given.

LXX. If, of two parallelograms, the sides including equal angles or including unequal but given angles have a given ratio to each other, these same parallelograms will also have a given ratio to each other.

Of two parallelograms *AB* and *EI*, let the sides including angles *Z* and *G*—whether they are equal or are unequal but given angles—have a given ratio to each other, that is, let the ratio *AG*:*EZ* be given and also the ratio *BG*: *ZI*. I say that the ratio *GD*:*ZT* is also given.

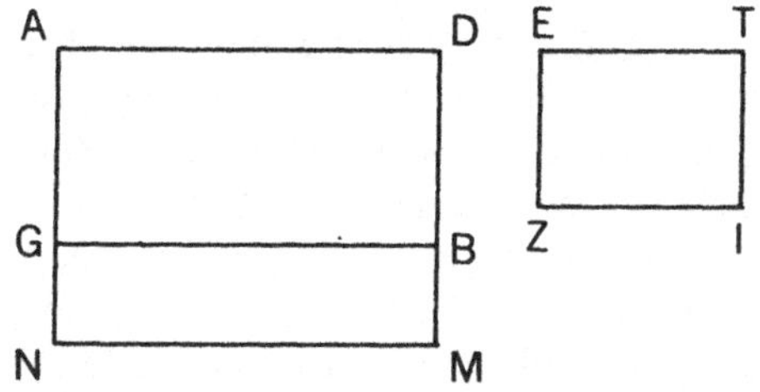

Let *GD* be equiangular with *ZT* and let there be applied to the straight line *GB* the rectangle *GM* equal to the parallelogram *ZT* and so posited that *AG* is in a straight line with *GN*. Then *DB* is also in a straight line with *BM*. And *BN* = *ZT*. But they are also equiangular. Therefore the sides of *BN* and *ZT* including the equal angles are reciprocally proportional. Therefore *GB*:*ZI* = *ZE*:*GN*. But the ratio *GB*:*ZI* is given. Therefore the ratio *EZ*:*GN* is also given. But the ratio *EZ*:*AG* is also given. Therefore the ratio *AG*:*GN* is given. Hence the ratio *GD*:*GM* is given. But *GM* = *ZT*. Therefore the ratio *GD*:*EI* is also given.

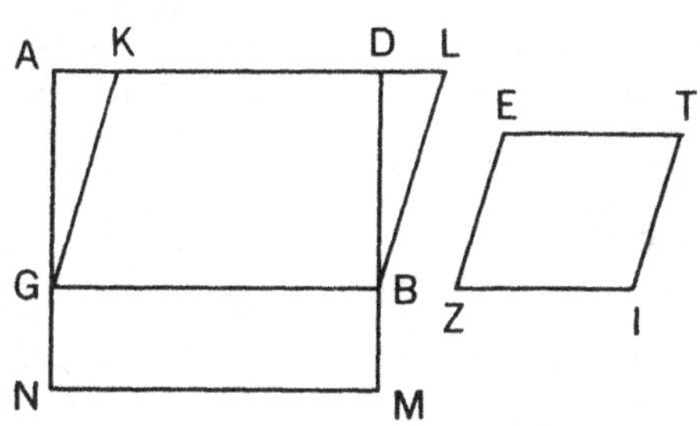

rectam et ad eius punctum *G* angulo *EZI* equalis angulus *BGK*, et compleatur *GL* parallelogrammum. Et quoniam datus est *AGB* angulus, et reliquus ergo *AGK* est datus. Est autem et *GAK* datus. Et reliquus ergo *AKG* (68v) datus est. Datus est ergo *AGK* trigonus specie./Proportio ergo est *AG* ad *GK* data. Sed et *AG* ad *EZ* proportio est data. Et *GK* ergo ad *EZ* proportio est data. Est autem et *GB* ad *ZI* porportio est data. Et est equalis *KGB* angulus angulo *EZI*. Proportio ergo *GL* ad *ZT* est data. Equale vero est *GL* ei quod est *GD*. Proportio ergo est et *GD* ad *ZT* data.

LXXI. Si duorum trigonorum circa equales angulos vel inequales quidem datos autem latera adinvicem proportionem habeant datam, eadem quoque adinvicem trigona proportionem habent datam.

Duo enim trigona *ABG* et *DET* circa equales angulos qui ad *A*, *D* vel circa inequales quidem datos autem latera adinvicem proportionem habeant datam et sit proportio quidem *AB* ad *ED* data et *AG* ad *DT*. Dico quod et *ABG* trigoni proportio est data ad *EDT* trigonum.

Compleantur enim *AI*, *DZ* parallelogramma. Quoniam ergo duorum parallelogrammorum *AI* et *DZ* circa equales angulos vel circa inequales quidem datos autem qui scilicet ad *A* et *D* latera adinvicem proportionem habent datam, et parallelogramma proportionem habent datam adinvicem. Proportio ergo *AI* ad *DZ* data. Et est parallelogrammi quidem *AI* medietas *ABG* trigonus. Eius vero quod est *DZ* trigonus *DET*. Porportio ergo *ABG* ad *DET* trigonum data.

1 eius *om. B* / EZI:ZI *B*
1–2 compleatur.... angulus *illeg. D*
3 est *tr. B post* datus[1] / Est *om. B* / autem: datus *P* / ergo *om. B*
3–4 datus[2].... AGK *illeg. D*
4 est[1] *om. B* / est[3] *om. B*
5–6 EZ[1].... da]ta[1] *illeg. D* / Et.... data[1] *om. B*
6 *post* data[2] *bis D* Est autem [et GB ad ZI proportio est data / Et .... K]GB *illeg. D*
7 GL: GI *B* / est[1] *om. B*
7–8 vero.... est[2] *illeg. D*
8 ergo *om. B* / ZT:ZD *P*
9 LXXI: 70[a] *B* / Si .... trigonorum *illeg. D*
10 quidem: quod *P* / autem .... adinvicem *illeg. D*
11 tri[gona .... datam *illeg. D*
12 vel circa *illeg. D* / *post* circa[2] *bis D* equales angulos qui ad A, D] vel circa
13–14 habeant.... qui]dem *illeg. D*
15 EDT:ED *B*
15–16 data.... Complean]tur *illeg. D*
16 Compleantur: Compleatur *BP* / AI, DZ: ad IDZ *P*
16–17 parallelogramma.... DZ *om. P*
17 circa[1]: contra *D*
17–18 angulos.... quidem *illeg. D*
18 autem: aut *B*
19 habent: habeant *DO* habeat *B* / datam[1].... habent *illeg. D*
20 quidem *om. D*
21 ABG.... DZ *illeg. D* / DET:DEZ *B* / Proportio ergo *bis D*
22 DET:ET *B*

Now let AB be not equiangular with ZT. On the straight line BG and at its point G let the angle BGK be constructed equal to the angle EZI and let the parallelogram GL be completed. Since the angle AGB is given [and the angle KGB also is given],* hence the remaining angle AGK is also given. But [the angle] GAK is also given. Therefore the remaining angle AKG is also given. Therefore the triangle AGK is given in species. Therefore the ratio $AG{:}GK$ is given. But the ratio $AG{:}EZ$ is also given. Therefore the ratio $GK{:}EZ$ is given. Now the ratio $GB{:}ZI$ is given. And the angle KGB is equal to the angle EZI. Therefore the ratio $GL{:}ZT$ is given. But $GL = GD$. Therefore the ratio $GD{:}ZT$ is also given.

LXXI. If, of two triangles, the sides including equal angles or including unequal but given angles have a given ratio to each other, these same triangles have a given ratio to each other.

Of the triangles ABG and DET, let the sides including the angles at A and D—whether they are equal or unequal but given angles—have a given ratio, and let the ratio $AB{:}ED$ be given, and also let the ratio $AG{:}DT$ be given. I say that the ratio $\Delta ABG{:}\Delta EDT$ is also given.

Let the parallelograms AI and DZ be completed. Since, of the two parallelograms AI and DZ, the sides including the angles at A and D—whether they are equal or unequal but given angles—have a given ratio to each other, the parallelograms also have a given ratio to each other. Therefore the ratio $AI{:}DZ$ is given. The triangle ABG is half of the parallelogram AI and the triangle DET is half of DZ. Therefore the ratio $\Delta ABG{:}\Delta DET$ is given.

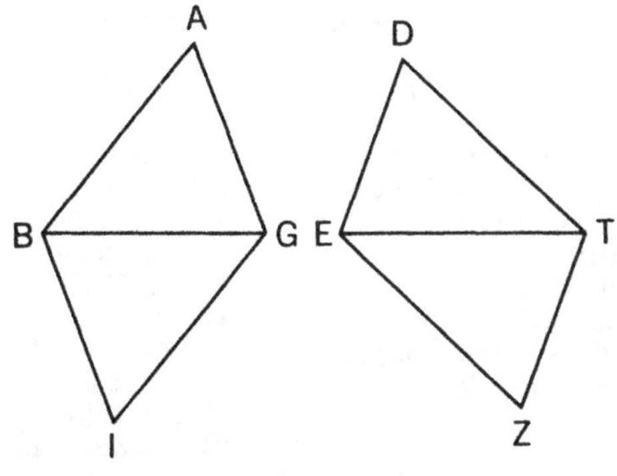

* This phrase is omitted in the body of the good Greek manuscript P, being only in the margins. Menge suggests that it should perhaps be omitted.

LXXII. Si duorum trigonorum bases in data proportione sint et super eas ducte ab angulis et he vel equales angulos faciant vel inequales quidem datos autem qui ad bases, et ipsa trigona adinvicem proportionem habebunt datam.

Sint duo trigona *ABG* et *DEZ* et protrahantur *AI* et *DT* vel equales angulos continentes qui videlicet sub *AIG* et *DTZ* vel inequales quidem datos autem, sit quoque proportio quidem *BG* ad *EZ* data, et *AI* ad *DT* data. Dico quod et *ABG* trigoni ad *DEZ* trigonum proportio est data.

Compleantur enim *KG*, *LZ* parallelogramma. Et quoniam *AIG* et *DTZ* anguli vel equales sunt vel inequales quidem dati autem, equalis vero angulus quidem *AIG* angulo *KBG*, at vero *DTZ* angulo *LEZ*, et qui ad *B* et *Z* ergo anguli vel equalis sunt vel inequales quidem dati autem. Et quoniam proportio est *AI* ad *DT* data, equalis vero *AI* quidem ei que est *KB*, at vero *DT* recte *LE*, proportio ergo est et *KB* ad *LE* data. Est vero et *BG* ad *EZ* proportio data. Et qui ad *B*, *E* puncta anguli vel equales sunt vel inequales quidem dati autem. Et parallelogrammi ergo *GK* ad *LZ* parallelogrammum proportio est data. Quare et *ABG* trigoni ad *DEZ* trigonum proportio est data.

LXXIII. Si duorum parallelogrammorum circum equales angulos vel inequales quidem datos autem latera sic se habeant ut sint sicut primi latus ad secundi latus ita reliquum secundi ad aliud quidem, habeat vero reliquum primi latus ad illud proportionem datam, ea

1 LXII *om. B* / Si . . . . trigon]orum *illeg. D*
2 ab . . . . equa]les *illeg. D* / he: hee *O*
3 autem . . . . ipsa *illeg. D*
4 proportionem *tr. B post* habebunt
5 Sint. . . .equa]les *illeg. D*
6 qui: quod *BP*
6–7 et. . . .quidem *illeg. D*
8 trigoni: trigonus *B*
8–9 Di[co. . . .Complean]tur *illeg. D*
10 anguli. . . .vero *illeg. D* / autem: aut *B*
11 ad B: al B *B*
11–12 DTZ. . . .qu]idem *illeg. D*
12 Z: E *B*
13 DT:AT *B*
13–14 AI[1]. . . .ergo *illeg. D* / equalis. . . data *om. B*
14–15 vero. . . .su]nt *illeg. D*
15 B:AB *B* / E:BE *B*
16–17 autem. . . .ad *illeg. D*
17 proportio[2] *om. B* / est[2] *bis P*
18 data:datum *B*
19 LXXIII: 62[a] *B* / circum: circa *B*
19–20 Si . . . . vel *illeg. D*
20 habeant: habent *B et corr. P ex* habent
20–21 datos . . . . latus[1] *illeg. D*
21 quidem: quid *BP*
21–22 la[tus[2] . . . . primi *illeg. D*
22–1(p.s.) ad . . . . adinvicem *illeg. D*

LXXII. If the bases of two triangles are in a given ratio, and if straight lines drawn to those bases from angles make either equal or unequal but given angles which are at the bases, those triangles will also have a given ratio to one another.

Let the two triangles be ABG and DEZ, and let AI and DT be drawn producing angles AIG and DTZ which are either equal or unequal but given angles. Let the ratio $BG{:}EZ$ be given and let the ratio $AI{:}DT$ also be given. I say that the ratio $\Delta ABG{:}\Delta DEZ$ is also given.

Let the parallelograms KG and LZ be completed. Since AIG and DTZ are either equal or unequal but given angles, $\angle AIG = \angle KBG$ and also $\angle DTZ = \angle LEZ$, hence the angles at B and Z are either equal or unequal but given. And since the ratio $AI{:}DT$ is given, $AI = KB$ and also $DT = LE$, therefore the ratio $KB{:}LE$ is also given. But the ratio $BG{:}EZ$ is also given. And the angles at points B and E are either equal or unequal but given. Therefore the ratio of the parallelograms $GK{:}LZ$ is also given. Hence the ratio $\Delta ABG{:}\ \Delta DEZ$ is also given.

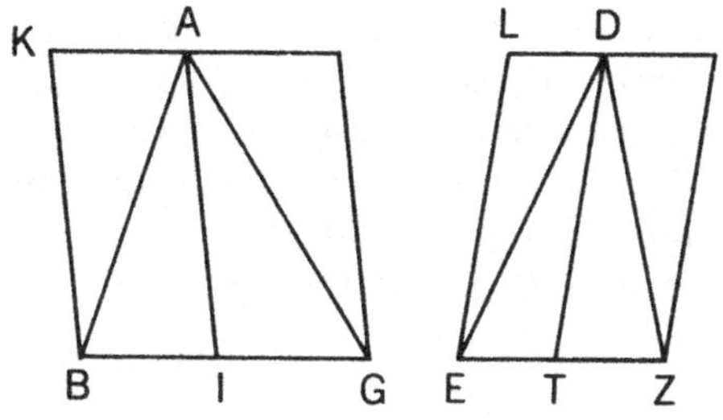

LXXIII. If, of two parallelograms, the sides including equal angles or including unequal but given angles are so related that one side of the first is to one side of the second as the remaining side of the second is to some other line, and if the remaining side of the first has a given ratio to that line, those parallelograms will also have a given ratio to each other.

Of two parallelograms AB and EI, let the sides including equal angles or including unequal but given angles at G and Z be to each other so that $GB{:}ZI = EZ{:}GK$ and let the ratio $AG{:}GK$ also be given. I say that the ratio of the parallelograms $GD{:}EI$ is also given.

quoque parallelogramma adinvicem proportionem habebunt datam.

Duorum enim parallelogrammorum *AB* et *EI* circum equales angulos vel inequales quidem datos autem qui videlicet ad *G*, *Z* latera ita se habeant adinvicem ut sint sicut *GB* ad *ZI* ita *EZ* ad *GK*, sit quoque *AG* ad *GK* proportio data. Dico quod et *GD* parallelogrammi ad *EI* parallelogrammum proportio est data.

Esto enim primum *AB* parallelogrammum parallelogrammo *EI* equiangulum et educatur secundum *GB* rectam *EI* parallelogrammo equale parallelogrammum *GT* et ponatur ut in directo sit *AG* ei que est *KG*. In directo ergo est et *DB* ei que est *BT*. Et quoniam equale est *GT* ei quod est *EI*, est autem et equiangulum, *GT*, *EI* ergo contraria patiuntur latera circa equales angulos. Est ergo sicut *GB* ad *ZI* ita *EZ* ad *GK*. Sicut vero *GB* ad *ZI* ita *EZ* ad quam *AG* proportionem habent datam, proportio ergo *AG* ad *GK* data. Quare et *AB* ad *GT*, hoc est, ad *EI* proportio est data.

Non sit autem equiangulum et constituatur ad rectam *GB* et ad eius (69r) punctum *G* angulo/EZI equalis angulus *BGL*, et compleatur *GM* parallelogrammum. Quoniam ergo datus est uterque horum *AGB* et *LGB*, et reliquus ergo *AGL* est datus. Datus est autem et *GAL* et reliquus ergo *GLA* datus est. Quare datus est et *AGL* trigonus specie. Proportio ergo *AG* ad *GL* data est. Quoniam est sicut *GB* ad *ZI* ita *EZ* ad quam *AG* proportionem habet datam, eius vero que est *AG* ad *GL* proportio est data, est ergo sicut *GB* ad *ZI* ita *ZE* ad *GL*. Et est equalis *LGB* angulus *EZI* angulo. Proportio ergo *GM* parallelogrammi ad *EI* parallelogrammum data. Equale vero *GM* ei quod est *GD*. Proportio ergo et *GD* ad *EI* data.

1–3 habebunt . . . . inequal]es *illeg. D*
2 enim: quidem *P* / circum: circa *B*
3 qui: que *B* / G, Z: D, G *B*
3–4 da[tos. . . .adinvic]em *illeg. D*
4 habeant: habent *B* / GB:DB *B*
4–5 sicut. . . .EI *illeg. D*
6–8 proportio. . . .secundum *illeg. D*
7 primum *tr. B post AB*
7–8 parallelogrammo. . . .parallelogrammo *om. P*
8–9 rectam. . . .AG *illeg. D* / equale parallelogrammum *om. B*
9–11 KG. . . .equales *illeg. D*
10 et *om. B* / equale *tr. B post* est^3 / GT:GD *B*
10–11 ei^2. . . .GT *om.(?) D*
12 GB^2:BG *B* / EZ^2:ET *B*
12–13 Sicut. . . .habent *illeg. D*
13 quam: qua *B*
13 GK:G *B*
14–15 et. . . .autem *illeg. D*
15–16 et^2. . .comple]atur *illeg. D*
16 EZI *correxi ex* EIZ *BOP*
17 horum *om. B* / LGB:LGG *BP*
17–18 Q[uoniam. . . .et^2 *illeg. D*
18 est^1 *tr. B post* datus^1 / est^2 *tr. B post* autem / ergo *om. P*
19–20 est^2. . . .GB *illeg. D*
20 est^1 *om. B* / *ante* Quoniam *scr. B* et / *post* AG *scr. et del. D* tt
20–22 EZ. . . .ita *om. B*
21–22 d[atam. . . .sicut *illeg. D*
22 *post* angulo *add. mg. infer. B notas illeg.*
22–23 angulus. . . .da]ta *illeg. D*
24 est *om. P* / EI:TI *P* / EI data *illeg. D*

In the first place, let the parallelogram *AB* be equiangular with the parallelogram *EI* and on the straight line *GB* let there be applied the parallelogram *GT* equal to the parallelogram *EI* and be so posited that *AG* is in a straight line with *KG*. Then *DB* is in a straight line with *BT*. Since *GT* = *EI* and they are also equiangular, the sides of *GT* and *EI* including the equal angles are reciprocally proportional. Therefore *GB*:*ZI* = *EZ*:*GK*. But *GB* is to *ZI* as *EZ* is to the line to which *AG* has a given ratio. Therefore the ratio *AG*:*GK* is given. Hence the ratio *AB*:*GT*, that is, *AB*:*EI* is also given.

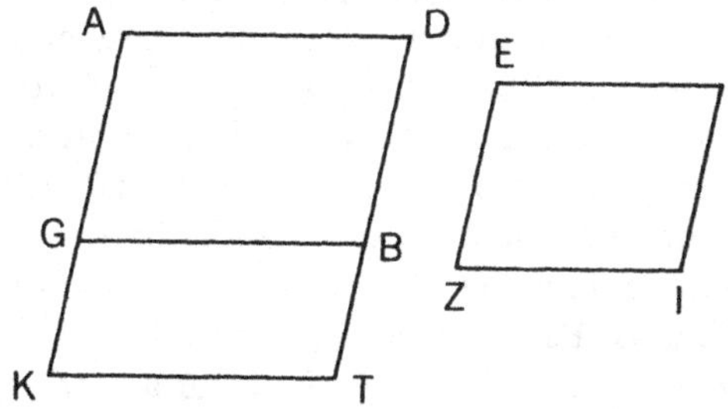

Now let the parallelogram *AB* be not equiangular to *EI* and let an angle *BGL* be constituted on the line *GB* and at its point *G*, equal to the angle *EZI* and let the parallelogram *GM* be completed. Since each of the angles *AGB* and *LGB* is given, the remaining angle *AGL* is also given.

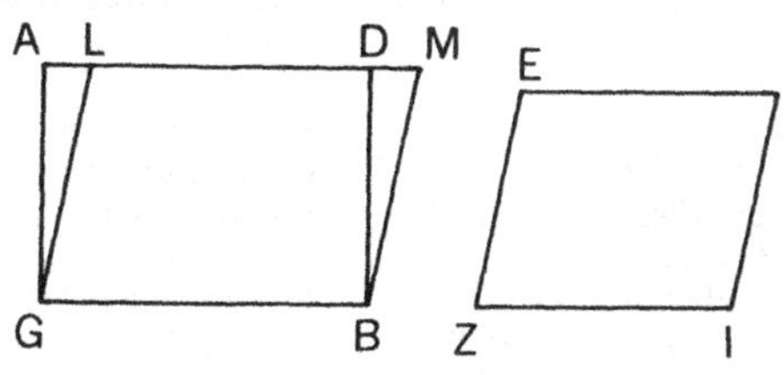

But the angle *GAL* and the remaining angle *GLA* are given. Hence the triangle *AGL* is given in species. Therefore the ratio *AG*:*GL* is given. Since *GB* is to *ZI* as *EZ* is to that straight line to which *AG* has a given ratio, so the ratio *AG*:*GL* is given. Therefore *GB*:*ZI* = *ZE*:*GL*. And $\angle LGB = \angle EZI$. Therefore the ratio of the parallelograms *GM*:*EI* is given. But *GM* is equal to *GD*. Therefore the ratio *GD*:*EI* is also given.

LXXIIII. Si duo parallelogramma proportionem habeant datam vel in equalibus angulus vel in inequalibus quidem datis autem, erit sicut primi latus ad secundi latus ita alterum secundi ad quod reliquum primi proportionem habet datam.

Duo enim parallelogramma *AB* et *EI* adinvicem proportionem habeant datam vel in equalibus angulis vel in inequalibus quidem datis autem qui videlicet ad *GZ*. Dico quod est sicut *GB* ad *ZI* ita *EZ* ad quam *AG* proportionem habet datam.

AB enim ei quod est *EI* vel equiangulum est vel non. Esto primum equiangulum et educatur secundum *GB* rectam parallelogrammo *EI* equale parallelogrammum *GT* et ponatur ut in directo sit *AG* ei que est GK. In directo ergo est et *DB* recte *BT*. Et quoniam proportio est *AB* ad *EI* data, equale vero *EI* ei quod est *GT*, proportio ergo *AB* ad *GT* data. Quare et *AG* ad *GK* proportio est data. Et quoniam equale est *GT* ei quod est *EI*, est autem et equiangulum, *GT*, *EI* ergo contrarie patiuntur latera que circum equales angulos. Est ergo sicut *GB* ad *ZI* ita *EZ* ad *GK*. Atqui *GK* ad *AG* proportio est data. Est ergo sicut *GB* ad *ZI* ita *EZ* ad quod *AG* proportionem habet datam.

Non sit autem equiangulum et constituatur ad *GB* rectam et ad eius punctum *G* angulo *EZI* equalis angulus *LGB* et compleatur *GM* parallelogrammum. Quoniam ergo proportio est *GD* ad *EI* data, equale vero *GD* ei quod est *GM*, proportio ergo *GM* ad *EI* data. Et est equalis angulus *LGB* angulo *EZI*. Est ergo sicut *GB* ad *ZI* ita *EZ* ad quod *GL* proportionem habet datam. At vero *GA* ad *GL* proportio est data. Est ergo sicut *GB* ad *ZI* ita *EZ* ad quod *AG* proportionem habet datam.

1 LXXIIII:73^a *B*
1–2 datam . . . . vel^2 *illeg. D*
3 primi . . . . latus *illeg. D* / latus^1 *om. B* / secundi^2 *om. B*
4–5 ha[bet . . . . habeant *illeg. D*
5 Duo: Due *B* / Duo enim: Si duo *P* / habeant: habeat *B*
6 *post* inequalibus *scr. et del. D* angulis
7 videlicet. . . .sicut *illeg. D*
9 EI. . . .pri]mum *illeg. D*
11 parallelogrammum. . . .sit *illeg. D* / que: quod *B*
12 est^1 *om. B*
12–13 AB. . . .quod *illeg. D*
13–14 Quare. . . .data *om. P*
14 data. . . .quod *illeg. D* / est^3 *om. P*
15 patiuntur: paciuntur *BP*
15–16 circum. . . .GB *illeg. D* / circum: circa *B*
17 ad^1: at *P* / sicut. . . .AG *illeg. D*
19–20 GB. . . .angulo *illeg. D*
21 est *tr. B post* data / equale *tr. B post* GD^2
21–22 GD^1. . . .quod *illeg. D*
22 *post* data *bis D* equale vero GD ei quod est GM [proportio ergo GM ad EI data. / est^2 *om. B*
22–23 Et. . . .LGB *illeg. D*
23 GB:GD *B*
24 habet. . . .est *illeg. D* / habet *tr. B post* datam

LXXIV. If two parallelograms either in equal angles or in unequal but given angles have a given ratio to each other, one side of the first is to one side of the second as the other side of the second will be to a straight line to which the remaining side of the first has a given ratio.

Let the two parallelograms *AB* and *EI* in equal angles or in unequal but given angles at *G* and *Z* have a given ratio to each other. I say that *GB* is to *ZI* as *EZ* is to the straight line to which *AG* has a given ratio.

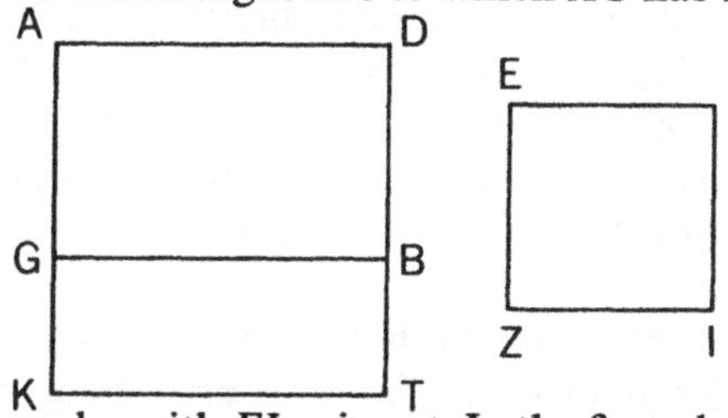

AB is either equiangular with *EI* or is not. In the first place, let it be equiangular. And on the straight line *GB* let the parallelogram *GT* be equal to the parallelogram *EI* and be so posited that *AG* is in a straight line with *GK*. Then *DB* is also in a straight line with *BT*. And since the ratio $AB{:}EI$ is given and $EI = GT$, the ratio $AB{:}GT$ is given. Hence the ratio $AG{:}GK$ is also given. And since $GT = EI$ and they are equiangular, hence the sides of *GT* and *EI* including the equal angles are reciprocally proportional. Therefore $GB{:}ZI = EZ{:}GK$. Now the ratio $GK{:}AG$ is given. Therefore *GB* is to *ZI* as *EZ* is to the straight line to which *AG* has a given ratio.

Now let *AB* be not equiangular with *EI* and let there be constructed on the straight line *GB* and at its point *G* the angle *LGB*, equal to the angle *EZI*, and let the parallelogram *GM* be completed. Since the ratio $GD{:}EI$ is given and $GD = GM$, therefore the ratio $GM{:}EI$ is given. And $\angle LGB = \angle EZI$. Therefore *GB* is to *ZI* as *EZ* is to the straight line to which *GL* has a given ratio. But the ratio $GA{:}GL$ is given. Therefore *GB* is to *ZI* as *EZ* is to the straight line to which *AG* has a given ratio.

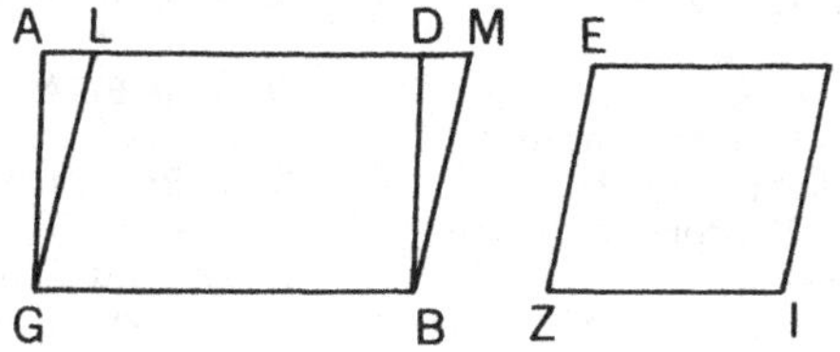

LXXV. Si duo trigona adinvicem proportionem habeant datam vel in equalibus angulis vel in inequalibus quidem datis autem, erit sicut primi latus ad secundi latus ita alterum secundi latus ad quod reliquum primi proportionem habet datam.

Sint duo trigona *ABG* et *DEZ* adinvicem proportionem habentia datam, et sint qui *A*, *D* anguli vel equales vel inequales quidem dati autem. Dico quod est sicut *AB* ad *DE* ita *DZ* ad quod *AG* proportionem habet datam.

Compleantur enim *AI* et *DT* parallelogramma. Et quoniam proportio est *ABG* trigoni ad *DEZ* trigonum data, proportio ergo parallelogrammi *AI* ad *DT* parallelogrammum data. Quoniam ergo duo parallelogramma *AI* et *DT* adinvicem proportionem habent datam vel in equalibus angulis vel in inequalibus quidem datis autem, est ergo sicut *AB* ad *DE* ita *DZ* ad quod *AG* proportionem habet datam.

LXXVI. Si a summitate trianguli dati specie super basem cathetus protrahatur, protracta ad basim proportionem habet datam.

Esto trigonus datus specie *ABG* et protrahatur ab *A* super *BG* cathetus *AD*. Dico quod proportio est *AD* ad *BG* data.

Quoniam enim datus est *ABG* trigonus specie, datus ergo est et *ABD* angulus. Est autem et BDA datus. Et reliquus ergo *BAD* est datus. Datus est ergo *ABD* trigonus specie. Proportio ergo est *BA* ad *AD* data. Sed et *AB* ad *BG* proportio est data. Et *AD* ergo ad *BG* proportio est data.

1 *propositionem LXXV tr. B post propositionem LXXVII*: / LXXV: 74[a] *B* / *ante* 74[a] *add. mg. B* hec propositio cum commento sequenti est 74[a] et intrat ibi / habeant: habent *O*
2 vel[1] . . . . angulis *illeg. D*
2–3 erit . . . . latus[2] *illeg. D*
3–4 quod reliquum *om. B*
4–5 reliquum . . . . ABG *illeg. D*
5–6 datam. . . . dati *illeg. D*
6 A, D *bis B*
7–8 ad[1]. . . . DT *illeg. D*
8 Compleantur: Compleatur *B*
9–10 ABG. . . . da]ta *illeg. D*
10 *post* data *scr. D* proportio / ergo *tr. D ante* Quoniam
10–12 parallelogramma. . . . vel *illeg. D*
12–13 au[tem. . . . datam *illeg. D*
14 LXXVI: 75[a] *B* / summitate: sumitate *P* / basem: bassem *P*
14–15 sum[mitate . . . . protracta *illeg. D*
15 protracta: protrahatur *P*
15–18 proportionem . . . . Quoniam *illeg. D*
17 ad *om. B*
18 ergo *tr. B post* est[2]
18–19 est[1]. . . . datus[2] *illeg. D*
19 *post* BDA *scr. B* angulus
19–21 est[1]. . . . data[2] *illeg. D*
20 ABD:AB *B* / Proportio *om. P* / est *om. B* / BA:AB *B* / AD:DA *O*
20–21 Sed. . . . data[1] *om. B*
21 proportio[1] *tr. P post* data[1]

LXXV. If two triangles in either equal angles or in unequal but given angles have a given ratio to each other, one side of the first is to one side of the second as the other [side] of the second is to the straight line to which the remaining [side] of the first has a given ratio.

Let the two triangles *ABG* and *DEZ* have a given ratio to each other. Let the angles at *A* and *D* be either equal or unequal but given. I say that *AB* is to *DE* as *DZ* is to the straight line to which *AG* has a given ratio.

Let the parallelograms *AI* and *DT* be completed. Since the ratio ΔABG:

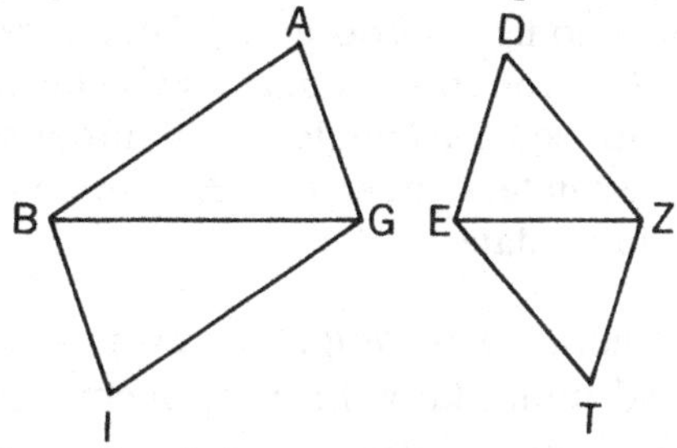

ΔDEZ is given, the ratio of the parallelograms *AI*:*DT* is given. Then since the two parallelograms *AI* and *DT* whether in equal angles or in unequal but given angles have a given ratio to each other, therefore *AB* is to *DE* as *DZ* is to the straight line to which *AG* has a given ratio.

LXXVI. If from the summit of a triangle given in species there is drawn to the base a perpendicular line, that line so drawn will have a given ratio to the base.

Let *ABG* be a triangle given in species and from *A* let the perpendicular *AD* be drawn to *BG*. I say that the ratio *AD*:*BG* is given.

Since the triangle *ABG* is given in species, the angle *ABD* is given. But the angle *BDA* is also given. Therefore the remaining angle *BAD* is also given. Therefore the triangle *ABD* is given in species. Therefore the ratio *BA*:*AD* is given. But the ratio *AB*:*BG* is also given. Therefore the ratio *AD*:*BG* is also given.

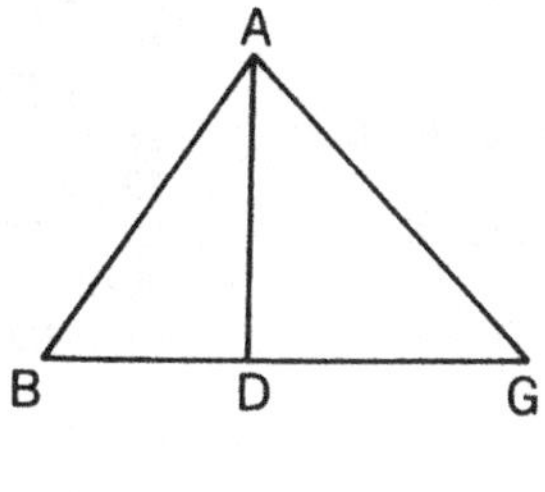

LXXVII. Si due figure date specie adinvicem proportionem habeant datam, et unumquodlibet unius figure latus ad quodcumque alterius proportionem habet datam.

Due enim figure *ABG* et *DEZ* date specie adinvicem proportionem
(69v) habeant datam. Dico quod unum/latus quodcumque figure *ABG* ad unum latus quodcumque figure *DEZ* proportionem habet datam.

Describantur enim ab *BG* et *EZ* tetragona *BI* et *ET*. Quoniam ergo ab eadem recta que est *BG* descripte sunt due figure que contingit date specie *ABG* et *BI*, proportio ergo figure *ABG* ad *BI* data. Et per eadem rursum etiam figure *DEZ* ad *ET* proportio est data. Quoniam ergo proportio est figure *ABG* ad *DEZ* data, sed *BAG* quidem ad *BI* proportio est data, *DEZ* vero ad *ET* proportio est data, et *BI* ergo ad *ET* proportio est data. Quare et *BG* ad *EZ* proportio est data.

LXXVIII. Si data figura ad rectangulum proportionem habeat datam et unum latus ad unum latus proportionem habeat datam, datum est rectangulum specie.

Data enim figura *AZB* ad rectangulum *GD* proportionem habeat datam, sitque proportio *ZB* ad *ED* data. Dico quod datum est *GD* specie.

Describatur enim ab *ZB* tetragonum *ZI* et educatur secundum *ED* parallelogrammo *ZI* equale parallelogrammum *EK* et ponatur ut in directo sit recta *GE* recte *ET*. In directo ergo est et *MD* recte *DK*. Et quoniam

1 LXXVII:76[a] *B*
1–2 due . . . . datam *illeg. D*
2 et *om. B*
2–3 unum[quodlibet . . . . datam *illeg. D*
4–5 enim. . . .figure *illeg. D*
5 ABG:AGB *B*
5–12 ad. . . .ET[1] *illeg. D*
7 Describantur: Describatur *B* / BG: G *P*
9–10 proportio. . . .data *om. B*
10 est[2] *om. B*
11 ABG:AGB *B* / BI correxi ex AI *BDOP*
12–13 est[2]. . . .data *illeg. D*
14 LXXVIII: 77[a] *B* / S[i . . . . ha]beat *illeg. D*
15 latus[2] *om. O*
15–16 latus[1] . . . . da]tum *illeg. D*
17 Data. . . .habeat *illeg. D*
18 datum: data *B* / *post* quod *add. mg. infer. B notas illeg.*
18–19 quod. . . .edu]catur *illeg. D*
19 enim *om. OP* / ZI:ZL *O*
20 ZI:ZL *DO*
20–21 et. . . .ET *illeg. D*
21 ergo *tr. B post* est

LXXVII. If two figures given in species have a given ratio to each other, the ratio will also be given for any side of the one figure to any side of the figure.

Let the two figures *ABG* and *DEZ* given in species have a given ratio to each other. I say that any side of figure *ABG* has a given ratio to any side of figure *DEZ*.

On *BG* and *EZ* let there be described the squares *BI* and *ET*. Since on the same straight line *BG* there are described any two figures given in species, *ABG* and *BI*, therefore the ratio of figure *ABG* to *BI* is given. By the same argument, the ratio of the figure *DEZ* to *ET* is also given. Since the ratio *ABG*:*DEZ* is given, the ratio *BAG*:*BI* is given and the ratio *DEZ*:*ET* is given, therefore the ratio *BI*:*ET* is also given. Hence the ratio *BG*:*EZ* is also given.

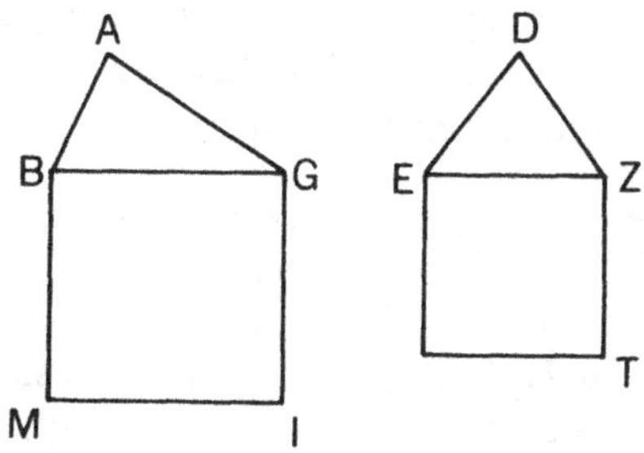

LXXVIII. If a given figure has a given ratio to a rectangle, and if one side has a given ratio to one [side], the rectangle is given in speices.

Let the given figure *AZB* have a given ratio to the rectangle *GD* and let the ratio *ZB*:*ED* be given. I say that *GD* is given in species.*

On *ZB* let the square *ZI* be described, and to *ED* let there be applied the parallelogram *EK* equal to the parallelogram *ZI* and it be so posited

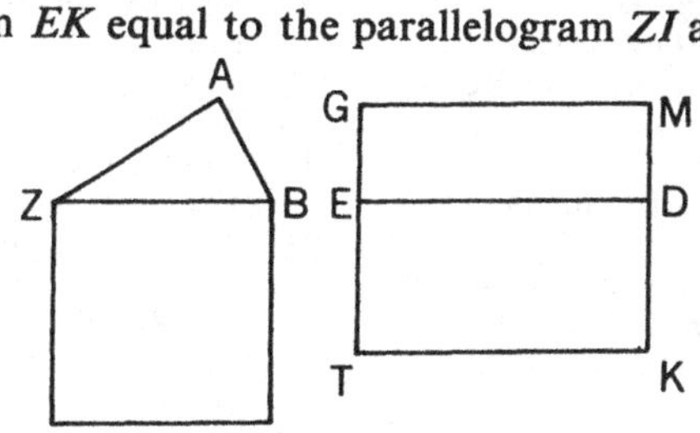

* The diagram here is in MS *P*. In MSS *O* and *D*, L and I are exchanged like 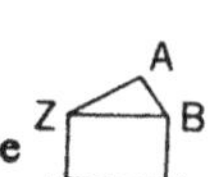

ab eadem recta *ZB* duo rectilinea que contingit data specie descripta sunt *AZB* et *ZI*, proportio est AZB ad *ZI* data. Atqui *AZB* ad *GD* proportio est data. Et *ZI* ergo ad *GD* proportio est data. Sed *ZI* ei quod est *EK* est equale. Et *GD* ergo ad *EK* proportio est data. Quare et *GE* ad *ET* proportio est data. Et quoniam equale est et equiangulum *ZI* ei quod est *EK*, est enim et rectangulum, contrarie ergo patiuntur eorum latera. Et est sicut *ZB* ad *ED* ita *ET* ad *ZL*, proportio vero supposita est recte *ZB* ad *ED* data. Proportio ergo et *ET* ad *ZL* data. Sed *ET* ad *GE* proportio est data. Et *GE* ergo ad *ZL* proportio est data. Equalis vero *ZL* recte *ZB*, tetragoni enim. Recte ergo *ZL* ad *ED* proportio est data, coniacet enim. Et *GE* ergo ad *ED* proportio data. Et est rectus qui ad *E* angulus. Datum est ergo *GD* specie.

LXXIX. Si duo trigona unum angulum uni angulo equalem habeant et ab equalibus angulis super bases catheti recte linee protrahantur, sicut vero primi trigoni basis ad cathetum ita alterius trigoni basis ad cathetum, equiangula erunt trigona.

Sint duo trigona *ABG* et *ZIT* angulos habentia ad *Z* et *B* equales et trahantur a punctis *Z* et *B* catheti *BD* et *ZK*, esto quoque sicut *AG* ad *BD* ita *TI* ad *KZ*. Dico quod equiangulus est *ABG* trigonus *TZI* trigono.

Circumscribatur enim *TIZ* trigono circulus, cuius sectio sit *TZI* et constituatur ad *TI* rectam et ad eius punctum *T* angulo *BAG* equalis *ITL*

1 re[cta. . . .specie *illeg. D* / *post* rectilinea *scr. P* parallilogramma
2 ZI[1] et ZI[2]:ZL *DO* / ad[1] *corr. D ex* et / Atqui: Sed *B*
2–3 da[ta. . . .Et *illeg. D*
3 ZI[1] et ZI[2]:ZL *DO* / est[2] *om. O*
3–4 ei. . . .equale: equale est EK *B*
4 equale. . . .ad[2] *illeg. D*
5 ZI:ZL *O* / EK:LK *B*
5–6 equian[gulum. . . .rectangulum *illeg. D*
6 patiuntur *correxi ex* paciuntur *BOP* atiuntur *D* / eorum *tr. B post* latera
7 ita. . . .recte *illeg. D* / ZL:ZI *O*
8 data[1]: datam *B*
8–9 Proportio. . . .data *om. B* / ET. . . ad *illeg. D*
9 ZL[2]:ZI *DO*
10 ZL:ZI *O* / enim[1]. . . .data *illeg. D*
11 ergo *om. P*
11–13 ad[2] . . . . un]um *illeg. D*
13 LXXIX *om. B* / *post* uni *scr. P* ab
14 equa[libus . . . . bases *illeg. D*
15 trigoni . . . . cathe]tum *illeg. D*
16–17 erunt . . . . trigona *illeg. D*
17 ZIT:ZTI *BP* / habentia: habencia *P et om. B*
18 puncti[s. . . .catheti *illeg. D* / quoque *om. B*
19 KZ:ZK *B* / Di[co. .TZI *illeg. D*
20 Circumscribatur: Circumscribantur *O*
20–21 sectio. . . .TI *illeg. D*

that *GE* is in a straight line with *ET*. Then *MD* is in a straight line with *DK*. Since on the same straight line *ZB* there are described any two rectilineal figures *AZB* and *ZI* given in species, the ratio *AZB*:*ZI* is given. Now the ratio *AZB*:*GD* is given. Therefore the ratio *ZI*:*GD* is also given. But *ZI* = *EK*. Therefore the ratio *GD*:*EK* is given. Hence the ratio *GE*:*ET* is given. Since *ZI* is equal to, and equiangular with, *EK*, for it is a rectangle, therefore their sides are reciprocally proportional. And *ZB*:*ED* = *ET*:*ZL*. But the ratio *ZB*:*ED* has been posited as given. Therefore the ratio *ET*:*ZL* is also given. But the ratio *ET*:*GE* is given. Therefore the ratio *GE*:*ZL* is given. But *ZL* = *ZB*, for *ZI* is a square. Therefore the ratio *ZL*:*ED* is given, for it is so posited. Therefore the ratio *GE*:*ED* is given. And the angle at *E* is a right angle. Therefore *GD* is given in species.

LXXIX. If two triangles have one angle equal to one angle, and if from the equal angles there are drawn straight lines perpendicular to the bases and if the base of the first triangle is to its perpendicular as the base of the other triangle is to its perpendicular, the triangles will be equiangular.

Let the two triangles *ABG* and *ZIT* have equal angles at *Z* and *B*, and from the points *Z* and *B* let there be drawn the perpendiculars *BD* and *ZK* and also let *AG*:*BD* = *TI*:*KZ*. I say that the triangle *ABG* is equiangular with the triangle *TZI*.

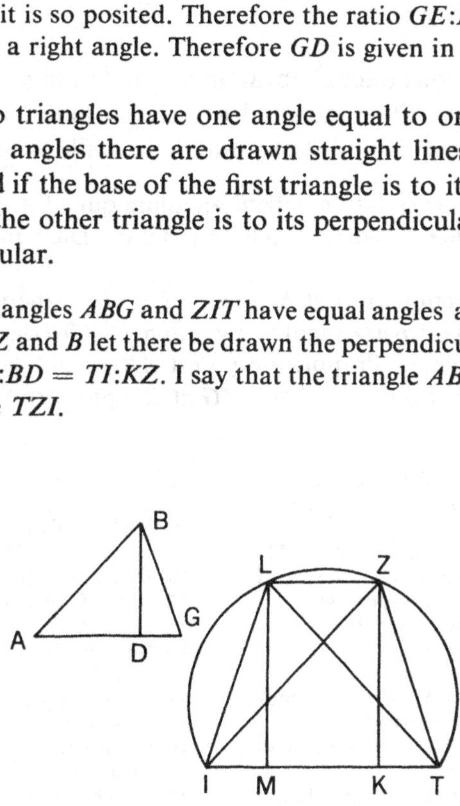

About the triangle *TIZ*, let a cricle be described and let its segment be *TZI*. Then, on the straight line *TI* and at its point *T*, let angle *ITL* be constructed equal to the angle *BAG* and let *Z* and *L*, and *L* and *I* be joined, and

angulus et coniungantur *Z, L* et *L, I* et trahatur cathetus *LM*. Quoniam ergo equalis est *BAD* angulus angulo *LTI*, est autem et *TLI* angulo *ABG* equalis, et reliquus ergo *BGA* reliquo *TIL* est equalis. Similis ergo est *ABG* trigonus *TIL* trigono. Et catheti tracte sunt *BD* et *LM*. Est ergo sicut *AG* ad *BD* ita *TI* ad *LM*. Erat autem sicut *AG* ad *BD* ita *TI* ad *ZK*. Equalis ergo est *LM* ei que est *ZK*. Est autem parallelos. Et *ZL* ergo recte *TI* parallela est. Equalis ergo est *ZLT* angulus *LTI* angulo. Sed et angulus quidem *LTI* angulo *BAG* equalis est et angulus *ZLT* angulo *ZIT*. Et angulos ergo *BAG* angulo *ZIT* est equalis. Est vero et *ABG* angulus angulo *TZI* equalis. Reliquus ergo *BGA* reliquo *ZTI* est equalis. Equiangulus ergo est *ABG* trigonus trigono *ZTI*.

LXXX. Si trigonus unum habeat angulum datum et quod sub datum angulum continentibus rectis ad ex reliquo latere tetragonum proportionem habeat datam, datus est trigonus ille specie.

Esto trigonus *ABG* datum habens angulum qui ad *A* et quod sub *BA* et *AG* ad quod ex *BG* proportionem habeat datam. Dico quod datus est *ABG* trigonus specie.

Protrahantur enim ab *A* et *B* super *BG* et *GA* catheti *BD* et *AE*. Quoniam ergo datus est *BAD* angulus, est autem et *ADB* datus, datus est ergo *ADB* trigonus specie. Proportio ergo est *AB* ad *BD* data. Quare et eius quod sub *BAG* ad illud quod sub *AG* et *BD* proportio est data. Ei vero

1 angulus: angulo *P* / Z, L et *om. B* / et[1]. . . .cathe]tus *illeg. D* / cathetus *om. B*
2 est[1] *om. B* / angulo[1] *om. B* / *de* TLI angulo *add. mg. O* ex XXVI tertii elementorum et ex ypothesi
2–3 et. . . .reliquus *illeg. D*
3 TIL:IL *B* / est[2] *om. B*
4–5 TIL. . . .Erat *illeg. D*
5 *supra* ita TI ad LM *add. O* ex penultima V[ti] elementorum
6 ergo[1] *tr. B post* est[1] / que: quod *B* / ergo. . . .TI *illeg. D*
6–7 parallela *correxi ex* paralilla *B* parallila *P* paralela *DO hic et ubique*
7 est[1] *om. B* / ergo est *om. B* / ZLT: LZT *BP et corr. O ex* LZT
8 et. . . .ZIT *om. D* / *supra* ZIT *add. O* catigmeni(?)
8–9 angulus. . . .BAG *illeg. D*
9 est *om. B*
9–10 equalis[1]. . . .Reliqu]us *illeg. D*
10 BGA *correxi ex* BAG *BDOP*/ZTI *correxi ex* ZTK *BDOP* / est[2] *om. B*
10–11 Equi[angulus. . . .ZTI *illeg. D*
12 LXXX *om. B* / Si . . . . datum *illeg. D*
13–14 continen[tibus. . . .proportionem *illeg. D*
14–15 trigo[nus . . . . sub *illeg. D*
15 BA:A *P*
16–18 proportionem. . . .Protra]hantur *illeg. D*
18 Protrahantur: Protrahatur *B* / et[1]: ad *B*
18–19 super. . . .datus[2] *illeg. D*
20 est *om. BP*
20–21 trigonus. . . .sub[2] *illeg. D*
21 quod[1] *bis P* / BAG:AG *B*
21–2(p.s.) proportio. . . .est *illeg. D*

let the perpendicular LM be drawn. Since $\angle BAD = \angle LTI$ and $\angle TLI = \angle ABG$, the remaining angle BGA is also equal to the remaining angle TIL. Hence the triangle ABG is similar to the triangle TIL. And the perpendiculars BD and LM are drawn. Therefore $AG{:}BD = TI{:}LM$. But $AG{:}BD = TI{:}ZK$. Therefore $LM = ZK$. But they are parallel. Therefore ZL is parallel to TI. Therefore $\angle ZLT = \angle LTI$. But $\angle LTI = \angle BAG$ and $\angle ZLT = \angle ZIT$. Therefore also $\angle BAG = \angle ZIT$. But also $\angle ABG = \angle TZI$. Therefore the remaining angle BGA is equal to the remaining angle ZTI. Therefore the triangle ABG is equiangular with the triangle ZTI.

LXXX. If a triangle has one given angle and if the rectangle of the straight lines containing the given angle has a given ratio to the square of the remaining [side], the triangle is given in species.

Let the triangle ABG have given angle at A and let the rectangle of BA and AG have a given ratio to the square of BG [*i.e.* let the ratio $(BA \times AG){:}\ BG^2$ be given]. I say that the triangle ABG is given in species.

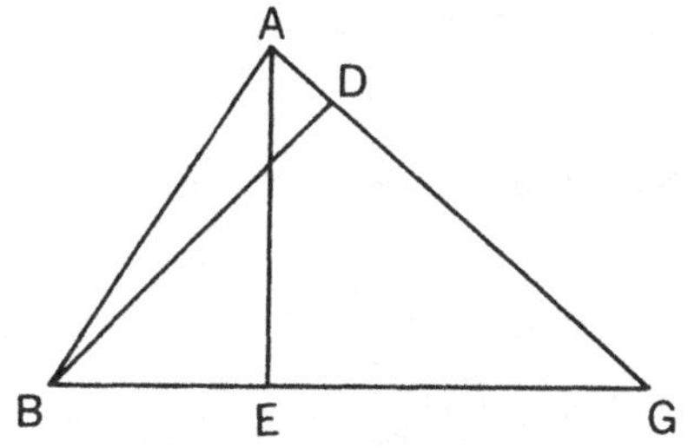

From the points A and B let the perpendiculars BD and AE be drawn to BG and GA. Since the angle BAD is given and the angle ADB is also given, the triangle ADB is given in species. Therefore the ratio $AB{:}BD$ is given. Hence the ratio $(BA \times AG){:}(AG \times BD)$ is given. But also $AG \times DB = BG \times AE$, for each of them is double of the triangle ABG. Therefore the ratio $(BA \times AG){:}(BG \times AE)$ is given. But the ratio $(BA \times AG){:}BG^2$ is given. Therefore the ratio $(BG \times AE){:}BG^2$ is also given and the ratio $BG{:}AE$ is given. Let the straight line ZI be given in position and magnitude and on ZI let there be described the segment of a circle ZTI having [inscribed in it] an angle equal to the angle BAG. Now the angle BAG is given. Therefore the angle in the segment ZTI is also given. Therefore the segment ZTI

quod sub *AG*, *DB* equale est quod sub *BG*, *AE*, utrumque enim eius dup- (70r) lum est trigoni *ABG*. Proportio ergo et eius quod sub *BAG* ad illud/quod sub *BG*, *AE* data. Eius vero quod sub *BA*, *AG* ad quod ex *BG* proportio est data. Et eius ergo quod sub *BG*, *AE* ad ex *BG* proportio est data et *BG* ad *AE* proportio est data. Adiaceat positione et magnitudine data recta *ZI* et describatur super *ZI* sectio *ZTI* datum habens angulum equalem angulo *BAG*. Datus autem *BAG* angulus. Datus ergo et in sectione *ZTI* angulus. Positione ergo est *ZTI* sectio. Protrahatur a puncto *I* recte *ZI* ad rectos recta *IK*. Positione ergo est recta *IK*. Et fiat sicut *BG* ad *AE* ita *ZI* ad *IK*. Proportio vero *BG* ad *AE* data. Proportio ergo et *ZI* ad *IK* data. Data vero *ZI*. Data ergo et *IK*, Sed et positione. Et est datum *I* punctum. Datum ergo et *K*. Protrahatur per *K* recte *ZI* parallela *KT*. Positione ergo est *KT*. Positione vero et *ZIT* sectio. Datus est *T* punctus. Coniungantur *ZT*, *TI* et trahatur cathetus *TL*. Data ergo est *TL*. Est autem *T* punctus datus et utrumque horum *Z*, *I*. Data ergo est unaqueque harum *TI*, *TZ*, *ZI* positione et magnitudine. Datus est ergo *ZIT* trigonus specie. Et quoniam est sicut *BG* ad *AE* ita *ZI* ad *IK*, equalis vero *IK* recte *TL*, est ergo sicut *BG* ad *AE* ita *ZI* ad *TL*. Et est equalis *BAG* angulus angulo *ZTI*. Equiangulus ergo est *ABG* trigonus *TZI* trigono. Datus vero *TZI* trigonus specie. Datus ergo et *ABG* trigonus specie.

Aliter

Esto trigonus *ABG* datum habens angulum qui videlicet ad *A*. Proportio

1 *de* eius *add. mg. O* trigoni scilicet quod apparet completo parallelogrammo continento sub BG et AE
2 est *om. B* / *post* trigoni *scr. BOP* / ergo *et del. O* / ABG:AGB *B*
2–3 ABG. . . .BA *illeg. D*
3 sub^{2} *om. B* / AG:EG *B*
3–4 quod^{1}. . . .BG^{2} *illeg. D*
4–6 data^{2}. . . .sectio *illeg. D*
6 ZI^{1}:ZL *B* / sectio: super *B*
6–7 da[tum. . . .ZTI *illeg. D*
8 angulus: angulo *B* / est *om. B* / *post* est *scr. et del. O* recta IK / *post* Protrahatur *scr. B* autem
8–9 an[gulus. . . .IK^{2} *illeg. D*
9–11 fi[at. . . .ZI *illeg. D*
10 *post* vero *scr. B* et
12 ergo. . . .KT^{1} *illeg. D* / Positione . . . .KT *om. B*
12–16 ergo^{2}. . . .et *illeg. D*
13 ZIT:ZTI *P* ZLT *B*
14 TL^{2}:GL *B* / *post* autem *scr. B* et
15 ergo *tr. B post* est
16 ZIT:TIZ *B*
16–17 Et. . . .vero *illeg. D*
17 vero *om. B*
17–18 ZI. . . .Equian]gulus *illeg. D*
18 *de* Equiangulus *add. mg. manu rec. O* per LXXIX
19 ergo *tr. post* est / ABG *om. B* / trigonus^{1} *tr. B post* TZI^{1} / TZI^{1}: ZI *B* / trigono *om. B* / TZI^{2}:TAG *B*
19–21 est. . . .Aliter *illeg. D*
20 trigonus *om. OP*
22–1(p.s) da[tum. . . .ad *illeg. D*

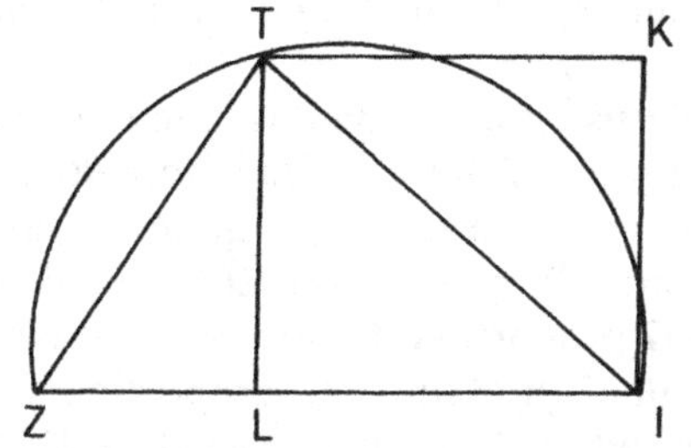

is given in position. From the point I let there be drawn the straight line IK at a right angle to ZI. Therefore the straight line IK is given in position. And let it be made that $BG{:}AE = ZI{:}IK$. But the ratio $BG{:}AE$ is given. Therefore the ratio $ZI{:}IK$ is also given. But ZI is given. Therefore IK is also given. But it is also given in position. And the point I is given. Therefore K is also given. Through K let KT be drawn parallel to ZI. Therefore KT is given in position. But the segment ZIT is also given in position. Therefore the point T is given. Let Z, T and T, I be joined and let the perpendicular TL be drawn. Therefore TL is given. But the point T is given and each of Z and I is given. Therefore each of TI, TZ and ZI is given in position and magnitude. Therefore the triangle ZIT is given in species. Since $BG{:}AE = ZI{:}IK$ and $IK = TL$, therefore $BG{:}AE = ZI{:}TL$. And $\angle BAG = \angle ZTI$. Therefore the triangle ABG is equiangular with the triangle TZI. Now the triangle TZI is given in species. Therefore ABG is also given in species.

Another proof

Let there be a triangle ABG having a given angle at A and let there be given the ratio of the rectangle of BA and AG to the square of BG. I say that the triangle ABG is given in species.

Since the angle BAG is given, that space by which $(BA + AG)^2$ is greater than BG^2 has a given ratio to the triangle ABG. Let the space by which $(BA + AG)^2$ is greater than BG^2 be D. Then the ratio $D{:}\Delta ABG$ is given. Now the ratio $\Delta ABG{:}(BE \times AG)$ is given, since the angle BAG is given. And the ratio $D{:}(BA \times AG)$ is given. But the ratio $(BA \times AG){:}BG^2$ is given. Therefore $D{:}BG^2$ is also given. Therefore, by composing, the ratio

vero sit eius quod sub *BA*, *AG* ad illud quod ex *BG* data. Dico quod datus est *ABG* trigonus specie.

Quoniam enim datus est *BAG* angulus, quo ergo maius est quod a coutraque *BAG* eo quod ab *BG*, illud spatium ad *ABG* trigonum proportionem habet datam. Quo autem maius est quod a coutraque *BAG* eo quod ab *BG* esto *D* spatium. Proportio ergo *D* spatii ad *ABG* trigonum data. Atqui *ABG* ad *BAG* proportio est data, eo quod datus sit *BAG* angulus. Et *D* ergo spatii ad illud quod sub *BAG* proportio est data. Eius vero quod sub *BAG* ad illud quod ex *BG* proportio est data. Et *D* ergo ad illud quod ex *BG* proportio est data. Et componenti proportio ergo *D* spatii cum eo quod ex *BG* ad illud quod ex *BG* est data. Sed *D* spatium cum eo quod ab *BG* quod a coutraque *BAG* est. Proportio ergo eius quod a coutraque *BA*, *AG* ad illud quod ex *BG* data. Quare et coutriusque *BAG* ad *BG* proportio est data. Et est datus *BAG* angulus. Datus est ergo *ABG* trigonus specie.

LXXXI. Si tres recte proportionales existentes tribus rectis proportionalibus existentibus extremas in data proportione habeant, et medias in data proportione habebunt. Et si extrema ad extremam proportionem habeat datam, et media ad mediam, et reliqua extrema ad reliquam extremam proportionem habebit datam.

Tres recte proportionales existentes *A*, *B*, *G* tribus rectis proportionalibus existentibus *D*, *E*, *Z* extremas in data proportione habeant. Et sit proportio *A* quidem ad *D* data, *G* vero ad *Z* proportio data. Dico quod et *B* ad *E* proportio est data.

Quoniam enim proportio est *A* quidem ad *D* data, *G* vero ad *Z* data, proportio ergo eius quod sub *A*, *G* ad illud quod sub *D*, *Z* data. Sed ei

2–3 ABG. . . .ergo *illeg. D*
3 *post* angulus *scr. et del. B* specie
4–5 illud. . . .datam *illeg. D*
5–6 eo quod ab BG *om. P*
6 BG. . . .ABG *illeg. D*
7 *supra* ABG *add. O* trigoni / *post* data *add. mg. O* ex LXVI[a]
7–8 data. . . .il]lud *illeg. D*
8–9 sub. . . .ad[1] *illeg. D*
9–10 Et. . . .data *om. B*
10–11 propor[tio[2]. . . .ex[2] *illeg. D*
11 BG[1] *bis B* / est *tr. P post* data
12 cout[raque. . . .coutra]que *illeg. D*
13 AG:G *BP*
13–14 B[AG . . . .specie *illeg. D*
14 *de* Datus *add. mg. manu rec. O* per XLV
15 LXXXI: 80[a] *B* / recte: linee *B*
15–16 rectis proportionalibus *illeg. D*
17 m[edias . . . . proportione *illeg. D*
18 habeat . . . . media *illeg. D*
18–19 extrema *om. B*
19 reliquam: reliquod *P* / ext[remam . . . .datam *illeg. D*
20 Tres: Trigonus *D*
20–21 proportion[alibus. . . .extre] mas *illeg. D*
22 vero. . . .B *illeg. D* / et *om. B*
24 enim: ergo *D* / est *om. B* / A *tr. B post* quidem
24–25 Z. . . .ad *illeg. D*

$(D + BG^2)$: BG^2 is given. But $D + BG^2 = (BA + AG)^2$. Therefore the ratio $(BA + AG)^2:BG^2$ is given. Hence the ratio $(BA + AG):BG$ is also given. And $\angle\ BAG$ is given. Therefore the triangle ABG is given in species.

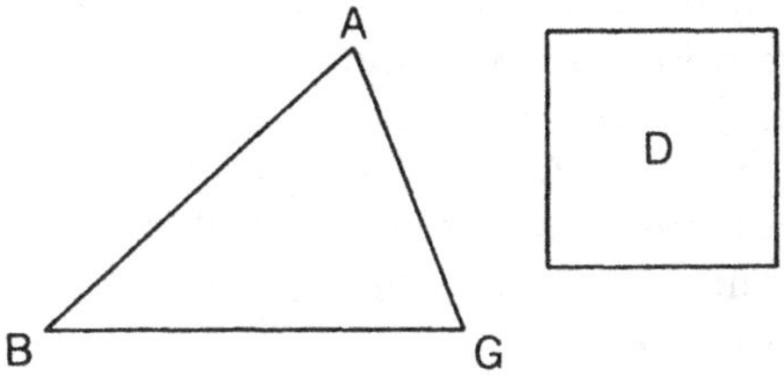

LXXXI. If three proportional straight lines have their extremes in a given ratio to the extremes of three [other] proportional straight lines, the means will also have a given ratio to the means. And if an extreme has a given ratio to an extreme and a mean to a mean, the remaining extreme will have a given ratio to the remaining extreme.

Let three proportional straight lines A, B and G have their extremes in a given ratio to the extremes of three [other] proportional straight lines D, E and Z and let the ratios $A:D$ and $G:Z$ be given. I say that the ratio $B:E$ is also given.

Since the ratios $A:D$ and $G:Z$ are given, the ratio $A \times G:D \times Z$ is given. But $A \times G = B^2$ and $D \times Z = E^2$. Therefore the ratio $B^2:E^2$ is given. Hence $B:E$ is also given.

Further, let the ratio $A:D$ be given and also let the ratio $B:E$ be given. I say that the ratio $G:Z$ is also given.

Since the ratio $A:D$ and $B:E$ are given, hence the ratio $B^2:E^2$ is also given. But $A \times G = B^2$ and $D \times Z = E^2$. Therefore the ratio $(A \times G):(D \times Z)$ is given. And the ratio of the one side A to the one side D is given. Therefore the ratio of the remaining side G to the remaining side Z is also given.

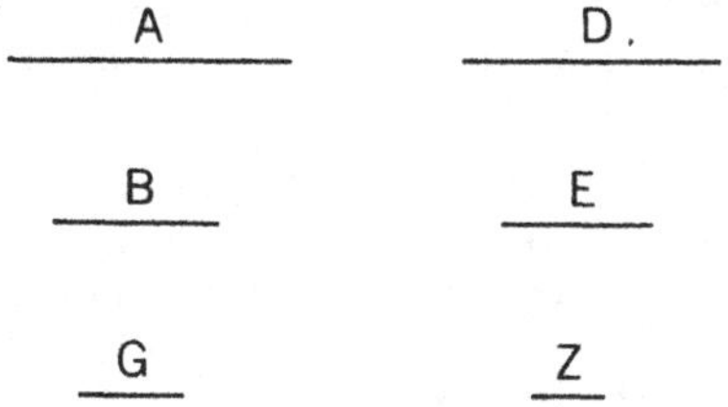

quod sub *A*, *G* equale est quod ab *B*, ei vero quod sub *D*, *Z* equale est quod ab *E*. Proportio ergo eius quod ab *B* ad illud quod ab *E* data. Quare et *B* ad *E* proportio est data.

Sit autem rursum *A* quidem ad *D* proportio data, *B* quoque ad *E* proportio sit data. Dico quod et *G* ad *Z* proportio est data.

Quoniam enim proportio est *A* quidem ad *D* et *B* quoque ad *E* data, proportio ergo et eius quod ab *B* ad illud quod ab *E* data. Sed ei quod ab *B* equale est quod sub *A*, *G*, ei vero quod ab *E* equale quod sub *D*, *Z*, proportio ergo est eius quod sub *A*, *G* ad illud quod sub *D*, *Z* data. Et uius lateris *A* ad unum latus *D* proportio est data. Et relique ergo *G* ad reliquam *Z* et proportio est data.

LXXXII. Si quatuor recte proportionales sint, erit sicut prima ad quam secunda proportionem habet datam ita tertia ad quam quarta proportionem habet datam.

Sint quatuor recte proportionales *A*, *B*, *G*, *D* sicut *A* ad *B* ita *G* ad *D*. Dico quod est sicut *A* ad quam *B* proportionem habet datam ita *G* ad quam *D* proportionem habet datam.

Esto enim ad quam *B* proportionem habet datam *E* et fiat sicut *B* ad
(70v) *E* ita *D* ad *Z*. Proportio vero *B* ad *E* data./Proportio ergo et *D* ad *Z* data. Et quoniam est sicut *A* ad *B* ita *G* ad *D*, est vero sicut *B* ad *E* ita *D* ad *Z*, per equum ergo sicut *A* ad *E* ita *G* ad *Z*. Et est quidem *E* ad quam

1 *post* G *bis D* ad illud quod sub D, Z data. [Sed ei quod sub A, G, / equale ....ab *illeg. D* / ei: si *B* / est *om. DOP*
2 E^1:BE *B*
2–3 ad....E *illeg. D*
3 proportio est *om. B*
4 rursum: rursus *P*
4–5 B....quod *illeg. D* / proportio *om. B*
5 proportio *tr. B post* est
6 quoque: quidem *B*
6–7 quoque....quod^3 *illeg. D*
7 proportio....data *om.(?) D*
8 B:E *B*
9 proportio....illud *illeg. D* / est *om. BP*
10 A *om. P* / relique: reliqua *B*
10–11 propor[tio....data *illeg. D*
11 reliquam: reliqua *B*
12 LXXXII: 81^a *B* / quatuor: *4 B hic et ubique*
12–13 proportion[ales .... secunda *illeg. D*
13–14 tertia .... datam *illeg. D*
15 quatuor: IIII^{or} *O* IIII *D*
15–16 proportion[ales....proportion]em *illeg. D*
16–18 ita....datam *om. B*
17–18 D....B^2 *illeg. D*
19 D^2:B *B*
19–20 Propor[tio^1....ita^1 *illeg. D*
20 est vero: esto *B* /
20–21 ad^2....quam *illeg. D*
21 equum: equam *B* / E^2:B *B*

LXXXII. If there are four proportional straight lines, the first will be to the straight line to which the second has a given ratio as the third is to the straight line to which the fourth has a given ratio.

Let A, B, G and D be four proportional straight lines, so that A is to B as G is to D. I say that A is to the line to which B has a given ratio as G is to the line to which D has a given ratio.

Let E be the line to which B has a given ratio and let it be made so that $B{:}E = D{:}Z$. Now the ratio $B{:}E$ is given. Therefore $D{:}Z$ is also given. Since $A{:}B = G{:}D$ and $B{:}E = D{:}Z$, therefore by equality $A{:}E = G{:}Z$. And E is that line to which B has a given ratio and Z is that line to which D has a given ratio. Therefore A is to the line to which B has a given ratio as G is to that line which D has a given ratio.

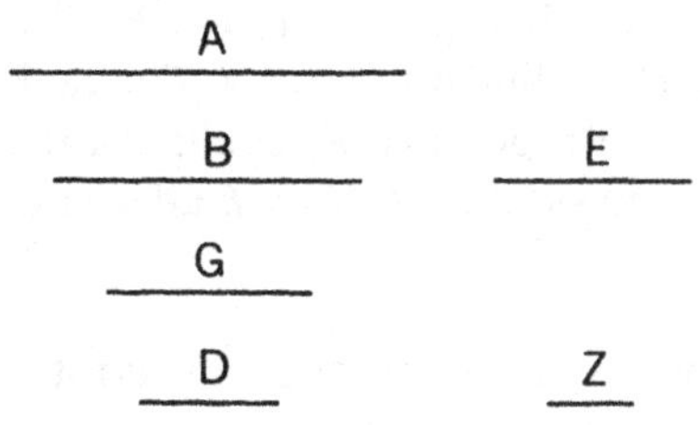

B proportionem habet datam, *Z* vero ad quam *D*. Est ergo sicut *A* ad quam *B* proportionem habet datam ita *G* ad quam *D* proportionem habet datam.

LXXXIII. Si quatuor recte ita se habeant adinvicem ut tribus simul sumptis ex eis quibuslibet et quarta eis assumpta proportionali, ad quam reliqua earum que e principio quatuor rectarum proportionem habet datam, proportionales fiant quatuor recte, erit sicut quarta ad tertiam ita secunda ad quam prima proportionem habet datam.

Sint quatuor recte *A*, *B*, *G*, *D* ita se habentes adinvicem ut tribus sumptis ex eis quibuslibet *A*, *B*, *G* et quarta eis assumpta *E*, ad quam *D* proportionem habet datam, proportionales sint *A*, *B*, *G*, *E* recte. Dico quod est sicut *D* ad *G* ita *B* ad quam *A* proportionem habet datam.

Quoniam enim sicut *A* ad *B* ita *G* ad *E*, quod ergo sub *A*, *E* equale ei quod sub *B*, *G*. Et quoniam proportio est *E* ad *D* data, proportio ergo et eius quod sub *A*, *D* ad illud quod sub *A*, *E* data. Ei vero quod *A*, *E* equale est quod sub *B*, *G*. Proportio ergo et eius quod sub *D*, *A* ad illud quod sub *B*, *G* data. Est ergo sicut *D* ad *G* ita *B* ad quam *A* proportionem habet datam.

LXXXIIII. Si due recte datum spatium contineant in dato angulo, altera vero altera dato maior sit, et utraque earum est data.

Due enim recte *AB*, *BG* datum spatium contineant *AG* in dato angulo *ABG*, sitque *GB* ea que est *BA* dato maior. Dico quod data est utraque *BA* et *BG*.

1 Est: Esto *B*
2 habet *om. B* / G:D *B* / D:G *B* / ita ....datam *illeg. D* / habet *om. B*
3 LXXXIII: 82^a *B* / ita *tr. B post* se
3–4 recte . . . . sumptis *illeg. D*
4–5 quibuslibet . . . . ea]rum *illeg. D*
5 reliqua: reliquam *P*
5–6 que . . . . fiant *illeg. D*
6 habet: habeat *B* / quarta: 4^a *B*
6–7 recte . . . . proportionem *illeg. D*
7–8 da[tam....B *illeg. D*
8 quatuor: IIII^{or} *O* IIII *D*
8–10 D....proportionem *illeg. D*
10 recte *om. B* / est *om. B*
10–17 datam....datam *illeg. D*
11 D: A *OP*
12 *post* ergo *scr. et del. O* quod
15 sub[2] *om. B*
16 D: A *OP*
17 *post* datam *add. O* ex LXXIIII
18 LXXXIIII: 83^a *B* / contineant: contineatur *B*
18–19 Si . . . . altera[1] *illeg. D*
19 altera[2] . . . . data *illeg. D*
20 contineant: continent *B*
20–21 da[to....Di]co *illeg. D*

LXXXIII. If four straight lines are so related that, any three of the lines being taken together with a fourth line which is assumed to be proportional to them and which has a given ratio to the remaining one of the original four lines, these [last] four lines are proportional, then the fourth will be to the third as the second is to that to which the first has a given ratio.

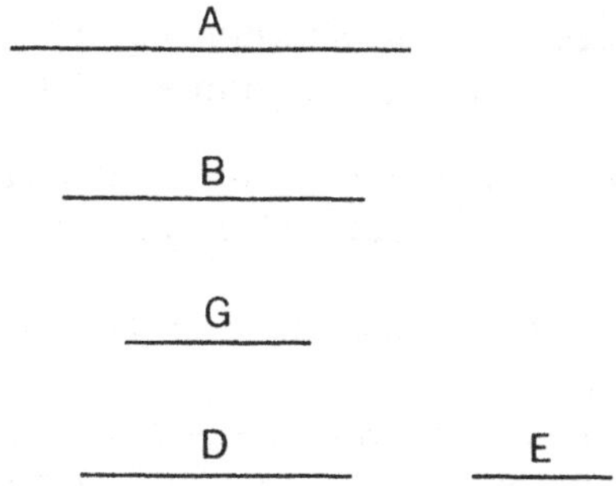

Let the four straight lines *A*, *B*, *G* and *D* be so related that, any three of them, *A*, *B* and *G*, being taken together with a fourth, *E*, which is assumed to be proportional to them and which has a given ratio to *D*, the straight lines *A*, *B*, *G* and *E* are proportional. I say that *D* is to *G* as *B* is to that to which *A* has a given ratio.

Since $A:B = G:E$, therefore $(A \times E) = (B \times G)$. Since the ratio $E:D$ is given, the ratio $(A \times D):(A \times E)$ is also given. But $(A \times E) = (B \times G)$. Therefore the ratio $(D \times A):(B \times G)$ is given. Therefore *D* is to *G* as *B* is to that to which *A* has a given ratio.

LXXXIV. If two straight lines contain a given space in a given angle and if the one is greater than the other by a given [line], each of the straight lines is also given.

Let two straight lines *AB* and *BG* contain a given space *AG* in a given angle *ABG* and let *GB* be greater than *BA* by a given [line]. I say that each of *BA* and *BG* is given.

Since *BG* is greater than *BA* by a given line, let that given line be *DG*. Then the remaining *DB* is equal to *AB*. And let *AD* be completed. Since $AB = DB$, the ratio $AB:BD$ is given. Now the angle *ABD* is given.

Quoniam enim *BG* quam *BA* dato maior est, esto datum illud *DG*. Reliqua ergo *DB* ei que est *AB* equalis est. Et compleatur *AD*. Et quoniam equalis est *AB* recte *BD*, proportio ergo *AB* ad *BD* data. Datus autem *ABD* angulus. Datum ergo est *AD* specie. Quoniam ergo *AG* datum secundum datam *DG* eductum est superhabundans figura data *AD*, data est ergo latitudo superhabundantie. Data ergo *BD*. Sed et *DG*. Et tota ergo *BG* data est. Est vero et *AB* data. Utraque ergo istarum *AB*, *BG* data est.

LXXXV. Si due recte datum spatium contineant in dato angulo et coutraque data, et unaqueque earum erit data.

Due enim recte *AB* et *BG* datum spatium contineant *AG* in dato angulo *ABG*, sitque coutraque *ABG* data. Dico quod et utraque data est *AB* scilicet et *BG*.

Protrahatur enim *GB* super *D* et sumatur *AB* recte recta *BD* equalis et per *D* punctum recte *BA* parallela trahatur *DE* et compleatur *AD*. Et quoniam equalis est recta *DB* recte *BA*, et est datus qui sub *ABD* angulus, quoniam et qui deinceps ei datus est, datum est ergo *EB* specie. Et quoniam data coutraque *ABG*, equalis vero *AB* recte *BD*, data ergo est recta *DG*. Quoniam ergo datum AG secundum datam *DG* eductum est dificiens specie dato *DA*, date sunt latitudines illius defectionis. Date ergo *AB* et *BD*. Sed et coutraque *ABG* data est. Data ergo utraque videlicet *AB* et *BG*.

1–2 quam....Reliqua *illeg. D* / Reliqua: Relinquum *B*
2–3 AD....AB^2 *illeg. D*
3 AB^1:GAB *B* / ABD:BD *B*
4 *de* AD specie *add. mg. O* per XLI sciens triangulis.
4–5 ergo^1....DG *illeg. D*
5 superhabundans: superabundans *B* / *de* data^2 *add. mg. manu rec. O* per LIX
5–6 data^2....Data *illeg. D*
6 superhabundantie *correxi ex* superhabundancie *BOP*
7 Est *om. D*
7–8 Utraque....spa]tium *illeg. D*
8 LXXXV *om. B*
9 et^2...data *illeg. D*
10 Due enim: Si due *D*
11 et *om. P*
11–13 AB....enim *illeg. D*
13 GB:BG *P* / recta BD: BD recte *B*
14 recte....et *illeg. D*
15 *post* est^2 *scr. et del. D* A
15–16 qui....qui *illeg. D*
17 ABG....BD *illeg. D* / ergo *tr. B post* est
18 recta *tr. B post* DG^1
18–19 est....su]nt *illeg. D*
20 ABG....videlicet *illeg. D* / *post* Data *scr. B* est / videlicet: videnorum(?) *B*

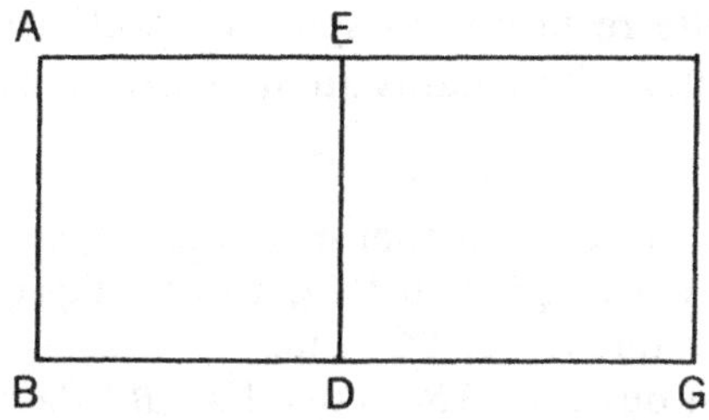

Therefore *AD* is given in species. Since the given space *AG* has been applied to the given straight line *DG*, exceeding it by the given figure *AD*, hence the length of the excess is given. Therefore *BD* is given. But *DG* is also given. Therefore the whole *BG* is also given. But *AB* is also given. Therefore each of *AB* and *BG* is given.

LXXXV. If two straight lines contain a given space in a given angle and the sum of these lines is given, each of these lines will be also given.

Let the straight lines *AB* and *BG* contain a given space *AG* in a given angle *ABG* and let the sum of these lines *AB* and *BG* be given. I say that each of *AB* and *BG* are also given.

Let *GB* be extended to *D*, and let *BD* be taken as equal to *AB*, and then through the point *D* let *DE* be drawn parallel to *BA* and let [the parallelogram] *AD* be completed. Since $DB = BA$ and the angle *ABD* is also given, for its adjacent angle is given, therefore *EB* is given in species. And since $AB + BG$ is given and $AB = BD$, hence *DG* is given. Since the given figure *AG* has been applied to *DG*, being deficient by the figure *DA* given in species, the dimensions of that deficient area are given. Therefore *AB* and *BD* are given. But $AB + BG$ is also given.* Therefore each of *AB* and *BG* is given.

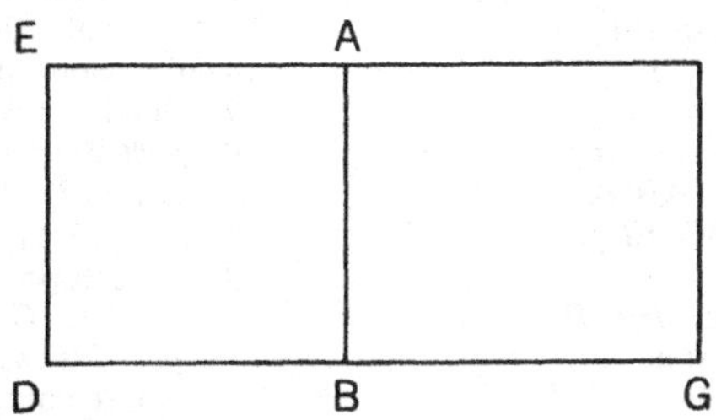

* Note that the Greek text in most manuscripts has an additional phrase: "therefore the remainder BG is given," but the Greek manuscript Vat. lacks the phrase. See Menge, p. 168.

LXXXVI. Si due recte datum spatium contineant in dato angulo, possitque una latera dato maius quam in proportione, utraque earum erit data.

Due enim recte *AB* et *BG* datum spatium contineant *AG* in dato angulo *ABG* et quod ab *GB* eo quod ab *BA* dato sit maius quam in proportione. Dico quod et utraque *AB* et *BG* data.

Quoniam enim quod ab *GB* eo quod ab *BA* dato maius est quam in proportione, auferatur datum illud quod sub *GBD*. Reliqui ergo quod sub *DGB* ad illud quod ab *AB* proportio est data. Et quoniam datum est quod sub *ABG*, est autem et quod sub *GBD* datum, proportio ergo est eius quod sub *ABG* ad illud quod sub *GBD* data. Sicut autem quod sub *ABG* ad illud
(71r) quod sub *GBD* ita *AB*/ ad *BD*. Quare et *AB* ad *BD* proportio est data. Quare et eius quod ab *AB* ad illud quod ab *BD* proportio est data. Eius vero quod ab *AB* ad illud quod sub *BGD* proportio est data. Et eius ergo quod sub *BGD* ad illud quod ab *BD* proportio est data. Quare et eius quod quater sub *BGD* ad illud quod ab *BD* proportio est data. Eius ergo quod quater sub *BGD* cum eo quod ab *BD* ad illud quod ab *BD* proportio est data. Sed illud quod quater sub *BGD* cum eo quod ab *BD* quod a coutraque est *BGD*. Proportio ergo et eius quod a coutraque *BGD* ad illud quod ab *BD* data. Quare et coutriusque *BGD* ad *BD* proportio est data. Et componenti ergo duorum *BG* ad *BD* proportio est data. Quare et unius *BG* ad *BD* proportio est data. Et *BG* ergo ad *BD* proportio est data. Sicut

1 LXXXVI *om. B* / Si . . . . datum *illeg. D*
2 latera . . . . quam *illeg. D* / latera: aliam *B*
4 Due. . . .spatium *illeg. D*
5–6 dato. . . .quod *illeg. D*
6 et[1] *om. B*
7–8 dato. . . .datum *illeg. D*
9 DGB:DG *B* / AB:BA *O*
9–10 AB. . . .ABG *illeg. D*
10 est[2] *om. B*
11 ABG. . . .Sicut *illeg. D*
12 Quare *tr. B post* et
12–13 AB[1]. . . .eius *illeg. D*
13–14 da[ta. . . .ergo *illeg. D*
14 BGD:DZE *B*
15 BGD:EZD *B*
15–16 est. . . .data *illeg. D*
16 BGD:EZD *B*
17 BGD:EZD *B*
17–18 cum. . . .BGD *illeg. D*
18 BGD:EZD *B* / *post* BD *scr. B* cum / a: ab *DOP*
18–20 contraque. . . .et *illeg. D*
19 BGD[1] *et* BGD[2]:EZD *B*
20 *post* BD[1] *add. mg. B* ex 2° euclidis / et *om. P* / BGD:EZD *B*
20–22 proportio. . . .B]G[1] *illeg. D*
21 Et componenti *om. B* / BG *correxi ex* DG *OP* DZ *B* / BD :BB *B* / *post* data *scr. B* proportio est data. Quare coutraque DZE excedit BD.
22 BG[1] *correxi ex* DG *O* GD *P* DZ *B*
22–1(p.s.) BG[2]. . . .illud *illeg. D*

LXXXVI. If two straight lines contain a given space in a given angle and the square of the one is greater than the square of the other by a given space as in ratio, each of those lines will be given.

Let the two straight lines *AB* and *BG* contain the given space *AG* in a given angle *ABG* and let GB^2 be greater than BA^2 by a given space as in ratio [*i.e.* the excess of BG^2 over a given space has a given ratio to BA^2]. I say that each of *AB* and *BG* is also given.

Since GB^2 is greater than BA^2 by a given space as in ratio [*i.e.* since the excess of GB^2 over a given space has a given ratio to BA^2], let the given space $GB \times BD$ be taken away. Then the ratio of the remainder $DG \times GB$ to AB^2 is given. Since $AB \times BG$ is given and $GB \times BD$ is also given, the ratio $(AB \times BG):(GB \times BD)$ is given. But $(AB \times BG):(GB \times BD) = AB:BD$. Hence the ratio $AB:BD$ is given. Hence the ratio $AB^2:BD^2$ is given. But $AB^2: (BG \times GD)$ is given. Therefore the ratio $(BG \times GD): BD^2$ is also given. Hence the ratio $(4BG \times GD):BD^2$ is also given. Therefore the ratio $(4BG \times GD + BD^2): BD^2$ is given. But $(4BG \times GD) + BD^2 = (BG + GD)^2$. Therefore the ratio $(BG + GD) :BD^2$ is also given. Hence $(BG + GD)^2:BD^2$ is given. Hence $(BG + GD):BD$ is given. And by composing, the ratio $2GB:BD$ is given. Hence the ratio of one *BG* to *BD* also is given. Therefore the ratio $BG:BD$ is given. But $GB:BD = (GB \times BD):BD^2$. Therefore the ratio $(GB \times BD):BD^2$ is given. But $GB \times BD$ is given. Therefore BD^2 is also given. Therefore *BD* is given. Hence *BG* is also given, for the ratio $BG:BD$ is given and *BD* is given. Therefore *BG* is also given.* And *AG* is given and the angle *ABG*** is given. Therefore *AB* is also given. Therefore each of *AB* and *BG* is given.***

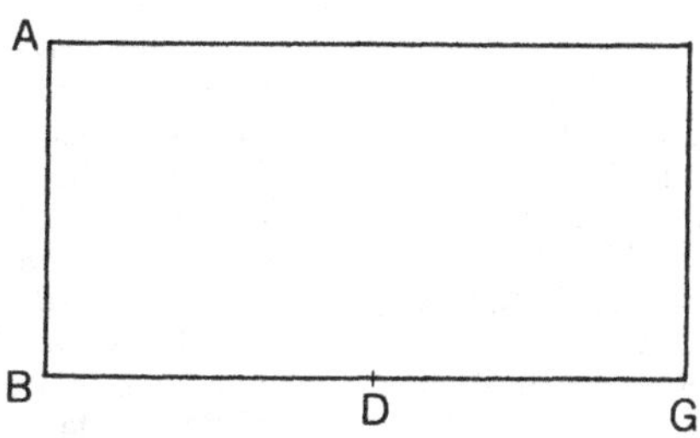

* This repetition is found in the Greek manuscripts P, Vat, and V. See Menge, p. 172.

** The most Greek manuscripts have *B* instead of *ABG*. MSS Vat. and v, like the manuscript used by the translator, have *ABG*. See Menge, p. 172.

*** The diagram in the Latin text

is replaced by the above in the Greek original.

vero *GB* ad *BD* ita quod sub *GBD* ad illud quod ab *BD*. Et eius ergo quod sub *GBD* ad illud quod ab *BD* proportio est data. Datum vero quod sub *GBD*. Datum ergo et quod ab *BD*. Data ergo *BD*. Quare et *BG* data est, nam *BG* ad *BD* proportio est data et data est *BD*. Data est ergo et *BG*. Et est datum *AG* et datus *ABG* angulus. Data ergo et *AB*. Utraque ergo *AB* et *BG* data est.

LXXXVII. Si in circulo dato magnitudine recta linea trahatur intercipiens sectionem recipientem angulum datum, data est protracta illa magnitudine.

In circulo enim dato magnitudine *ABG* protrahatur recta *AG* intercipiens sectionem *AEG* recipientem angulum datum. Dico quod *AG* data est magnitudine.

Sumatur enim circuli centrum *D* et coniuncta *AD* protrahatur super *E* et coniungantur *G*, *E*. Datus est ergo *AGE* angulus, rectus enim est autem. Et *AEG* datus. Et reliquus ergo *GAE* datus est. Datus est ergo *AGE* trigonus specie. Proportio ergo est *EA* ad *AG* data. Data vero *EA* magnitudine, quoniam et circulus est datus magnitudine. Data ergo est *AG* magnitudine.

LXXXVIII. Si in circulo dato magnitudine recta linea trahatur data magnitudine, intercipit sectionem recipientem angulum datum.

In circulo enim dato magnitudine *ABG* recta linea trahatur *AG* data magnitudine. Dico quod intercipit sectionem recipientem angulum datum.

1 GBD:BGD *B*
2 GBD:BGD *B*
2–4 Da[tum....ad *illeg. D*
3 *post* ergo *scr. B* quod
4 BG[1]: GB *OP* / BD[2]:DB *OP*
4–5 et[1]....Utraque *illeg. D*
5 *post* datus *scr. B* est / *post* AB[1] *add. mg. O* per LVII
7 LXXXVII *om. B* / recta *tr. B post* linea
7–8 Si . . . . intercipie]ns *illeg. D*
8 recipientem: recipiente *B*
8–10 recipientem . . . . In *illeg. D*
10 enim *illeg. D*
10–11 magnitudine....da]tum *illeg. D*
11 sectionem *tr. O post* AEG / recipientem: recipiunt *B*
11–13 quod....E *illeg. D*
14 *de* angulus *add. mg. manu rec. O* per XXX 3[1] libri euclidis
14–15 G....Datus *illeg. D*
15 datus[1] *om. P*
16 est *om. B* / vero: ergo *O*
17 datus *om. BP* / ergo *tr. B post* est
15–17 AG[E....datus *illeg. D*
17–18 Data....magnitudine *illeg. D*
19 LXXXVIII *om. B*
19–21 S[i....AG *illeg. D*
21 ABG:ABGD *BP et corr. O ex* ABGD
22–1(p.s.) sectionem....D *illeg. D*

LXXXVII. If in a circle given in magnitude there is drawn a straight line intercepting a segment which contains a given angle, that line so drawn is given in magnitude.

In the circle *ABG* given in magnitude, let there be drawn a straight line *AG* intercepting the segment *AEG* containing a given angle. I say that *AG* is given in magnitude.

Let *D* be taken as the center of the circle and let *AD*, when joined, be extended to *E* and let *G* and *E* be joined. Therefore the angle *AGE* is given, for it is a right angle. And [the angle] *AEG* is given. Therefore the remaining angle *GAE* is given. Therefore the triangle *AGE* is given in species. Therefore the ratio *EA*:*AG* is given. But *EA* is given in magnitude, since the circle is also given in magnitude. Therefore *AG* is given in magnitude.

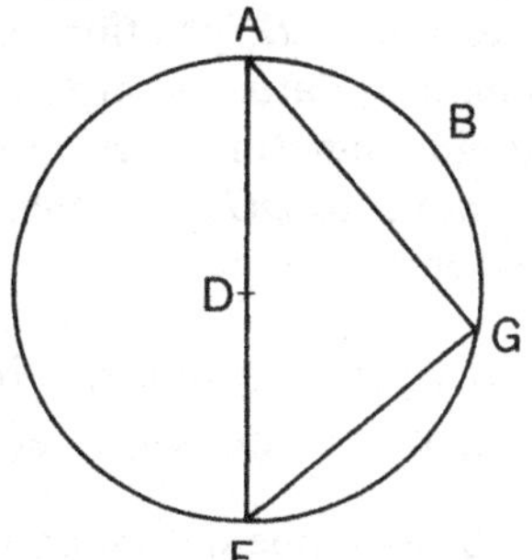

LXXXVIII. If in a circle given in magnitude there is drawn a straight line given in magnitude, that line intercepts a segment containing a given angle.

In the circle *ABG* given in magnitude, let there be drawn a straight line *AG* given in magnitude. I say that the line intercepts a segment containing a given angle

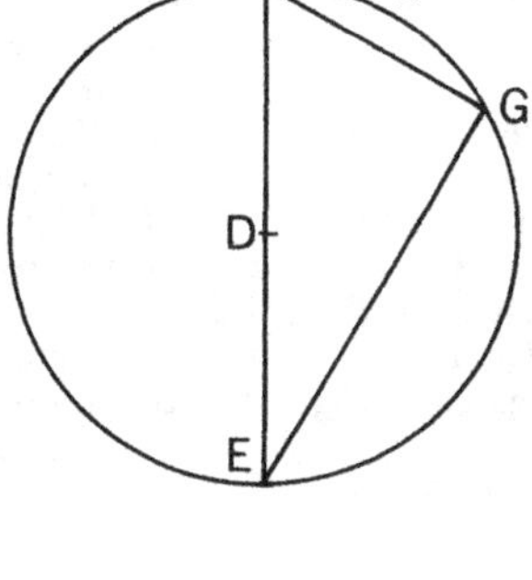

Sumatur enim centrum circuli *D* et coniugata *AD* protrahatur super *E* et coniungantur, *G*, *E*. Quoniam ergo data est utraque videlicet *EA* et *AG*, proportio ergo *EA* ad *AG* data. Et est rectus *AGE* angulus. Datus est ergo *AGE* trigonus specie. Datus est ergo et *AEG* angulus.

LXXXIX. Si circuli dati positione in periferia datus punctus sumatur ab eo vero ad circuli periferiam frangatur recta datum angulum faciens, datus est alter terminus illius fracte.

Circuli enim potitione dati *ABG* sumatur super periferiam datus punctus *B*, a puncto vero *B* frangatur recta *BAG* datum faciens angulum *BAG*. Dico quod datus est *G* punctus.

Sumatur enim centrum circuli *D* et coniungantur *BD*, *DG*. Et quoniam datus est uterque punctorum *B*, *D*, positione ergo est recta *BD*. Et quoniam datus est *BAG* angulus, datus est ergo et *BDG*. Quoniam ergo ad positione rectam et ad eius punctum *D* recta linea tracta est *DG* datum faciens angulum *BDG*, data ergo *DG* positione. Positione vero datus est et *ABG* circulus. Datus est ergo *G* punctus.

XC. Si a dato puncto positione datum circulum contingens recta trahatur, data est ea protracta positione et magnitudine.

A dato enim puncto *G* positione datum punctum circuli *AB* recta contingens trahatur *GA*. Dico quod *GA* recta data est positione et magnitudine

1 coniugata: coniungata *P* coniugatam *B*
2 coniunga[ntur. . . .EA *illeg. D*
3–4 data. . . .angulus *illeg. D*
4 Datus: Datum *B*
5 LXXXIX: 88[a] *B* / *ante* circuli *scr. B* in
5–6 periferia . . . . circuli *illeg. D*
6 circuli *om. B*
6–7 ang[ulum . . . . fracte *illeg. D*
8–9 punctus. . . .recta *illeg. D*
10–11 punctus. . . .coniungan]tur *illeg. D*
12 punctorum *om. B* / D:B *B*
12–13 recta. . . .est[2] *illeg. D*
13 est[2] *om. O*
14 positione: positionem *BP*
14–15 recta. . . .faciens *illeg. D*
15 data: datus *B* / datus est et *om. B*
16 Datus. . . .punctus *illeg. D* / *post* punctus *scr. B* datus est
17 XC: 89[a] *B*
17–18 rec[ta. . . .pro]tracta *illeg. D*
19 positione. . . .recta *illeg. D* / *post* recta *scr. B* linea
20 Dico. . . .positio]ne *illeg. D* / GA recta *tr. B post* magnitudine *ln. 20–21*

Let *D* be taken as the center of the circle and let *AD*, when joined, be extended to *E* and let *G* and *E* be joined. Since each of *EA* and *AG* is given, the ratio *EA*:*AG* is given. And *AGE* is a right angle. Therefore the triangle *AGE* is given in species. Therefore the angle *AEG* is also given.

LXXXIX. If on the circumference of a circle given in position there be taken a given point and if from the point a straight line is bent at the circumference of the circle making a given angle, the other extremity of that bent line is given.

On the circumference of the circle *ABG* given in position let there be taken a given point *B* and from the point *B* let the straight line *BAG* be inflected making a given angle *BAG*. I say that the point *G* is given.

Let *D* be taken as the center of the circle and let *BD* and *DG* be joined. Since each of the points *B* and *D* is given, the straight line *BD* is given in position. Since the angle *BAG* is given, *BDG* is also given. Since to the straight line [given] in position and at the point *D* there is drawn the straight line *DG* making a given angle *BDG*, therefore *DG* is given in position. Now the circle *ABG* is also given in position. Therefore the point *G* is given.

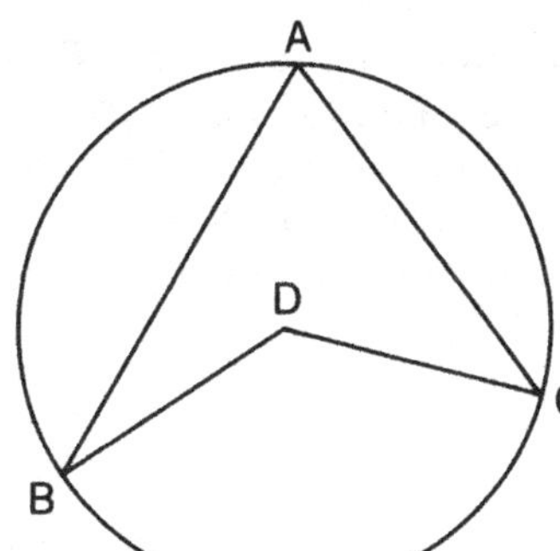

XC. If from a given point there is drawn a straight line touching a circle given in position, that line so drawn is given in position and magnitude.

From a given point *G* let there be drawn a straight line *GA* touching a circle *AB* given in position.* I say that the straight line *GA* is given in position and magnitude.

* The Latin translation "datum punctum circuli" is not right. It should be "positione datum circulum." See Menge, p. 177.

Sumatur centrum circuli *D* et coniungantur *DA*, *DG*. Et quoniam datus est uterque punctorum *D*, *G*, data ergo est recta *DG*. Et est rectus *DAG* angulus. Ergo super *DG* descriptus semicirculis veniet per *A*. Veniat et esto *DAG*. Positione ergo *DAG*. Positione vero *AB* circulus. Datus est *A*. Sed et *G* datus est. Data ergo est *AG* positione et magnitudine.

XCI. Si circulo dato positione sumatur aliquis punctus extra datus, a puncto vero in circulum protrahatur quedam recta, sub ea ducta
(71v) et ea que inter punctum/periferiamque convexam est recta contentum orthogonium datum est.

Circulo enim dato positione *ABG* sumatur punctus extra *D*, a puncto vero *D* protrahatur recta *BD* secans circulum. Dico quoniam datum est quod sub *BDG*.

Trahatur a puncto *D* circulum *ABG* contingens recta *AD*. Data ergo est *AD* positione et magnitudine. Quoniam ergo data est *AD*, datum ergo est et quod ab *AD*. Et est equale ei quod sub *BDG*. Datum ergo quod sub *BDG*.

Aliter

Sumatur centrum circuli *E* et coniuncta *DE* protrahatur super *A*. Et quoniam datus est uterque punctorum *E*, *D*, data ergo *ED*. Positione datus autem et *AB* circulus. Datus ergo uterque punctorum *A*, *Z*. Est vero et *D* datus. Data ergo utraque *AD* et *ZD*. Datum ergo est quod sub *AD*, *DZ*. Et est equale ei quod sub *BDG*. Datum ergo et quod sub *BDG*.

1 circuli *tr. B post* D / DA, DG: DAG *B*
1–2 Et....G *illeg. D*
2 ergo *tr. B post* est
3–4 semicirculis....esto *illeg. D*
5 Da[ta....magnitudine *illeg. D*
6 XCI: 90^a *B*
6–7 extra . . . . puncto *illeg. D*
7–8 duc[ta . . . . punctum *illeg. D*
9 orthogonium: ortogonium *B hic et ubique*
9–10 orthogon[ium . . . . su]matur *illeg. D*
10 *post* sumatur *scr. B* enim
11–12 secans....sub *illeg. D*
13 a: autem *B* / contingens: continges *P* / ergo *tr. B post* est
13–14 D[ata....magnitudine *illeg. D*
14 ergo^2 *tr. B post* est^2
15–18 sub....Su]matur *illeg. D*
18–19 Et....Positio]ne *illeg. D*
19 data: est *B* / *post* ED *bis B* est ergo ED
20 autem *correxi ex* ergo *BDOP*
20–21 A....Datum *illeg. D*
21 ergo^2 *tr. B post* est / *post* est *scr. O* et
22 BD[G....BDG *illeg. D* / BDG^2: DBG *B*

Let D be taken as the center of the circle and let DA and DG be joined. Since each of the points D and G is given, the straight line DG is given. And the angle DAG is a right angle. Therefore the semicircle described on DG will pass through [the point] A. Let it pass and let the semicircle be DAG. Then [the semicircle] DAG is given in position. But the circle AB is given in position. [Therefore] A is given. But G is also given. Therefore AG is given in position and magnitude.

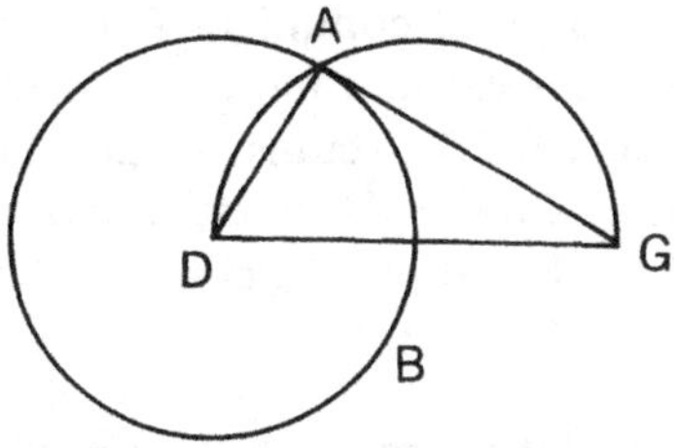

XCI. If outside a circle given in position there is taken some given point and from the point there is drawn a certain straight line into the circle, the rectangle comprised by the straight line so drawn and the straight line which is between the point and the convex circumference is given.

Outside the circle ABG given in position let there be taken the [given] point D and from the point D let there be drawn the straight line BD cutting the circle. I say that $BD \times DG$ is given.

From the point D let the straight line AD be drawn, touching the circle ABG. Then AD is given in position and magnitude. Since AD is given, AD^2 is also given. And $AD^2 = BD \times DG$. Therefore $BD \times DG$ is given.

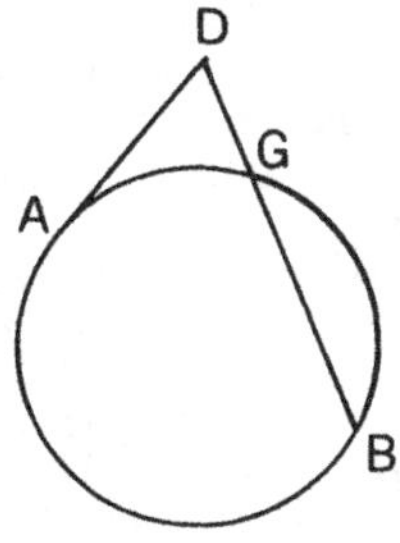

Another proof

Let E be taken as the center of the circle and let DE, when joined, be extended to A. Since each of the points E and D is given, the straight line ED is given. But the circle AB is also given in position. Therefore each of the points A and Z is given. But D is also given. Therefore each of AD

XCII. Si circulo dato positione sumatur punctus intus datus, per eum vero punctum protrahatur recta in circulum, sub eius protracte sectionibus contentum orthogonium datum est.

Circulo enim dato positione *BG* sumatur punctus intus videlicet *A* datus et per eum punctum *A* protrahatur recta *BG*. Dico quoniam datum est quod sub *BAG*.

Sumatur centrum circuli *D* et coniuncta *DA* protrahatur super puncta *E*, *Z*. Quoniam ergo datus est uterque punctorum *D*, *A*, positione ergo est *DA*. Positione vero et *GZB* circulus. Datus ⟨ergo⟩ est uterque punctorum *Z*, *E*. Est autem et *A* datus. Data est ergo utraque *ZA* et *AE*. Datum est ergo quod sub *ZAE*. Et est equale ei quod sub *BAG*. Datum ergo et quod sub *BAG*.

XCIII. Si in circulum datum magnitudine recta linea trahatur intercipiens portionem recipientem angulum datum, et in portione angulus in duo equa dividatur, coutreque datum angulum continentes ad in duo dividentem angulum proportionem habebunt datam, et quod sub coutraque datum angulum continentium rectarum atque deorsum intercepta ab in duo equa dividente angulum ad circumferentiam datum erit.

In circulo enim dato magnitudine *ABG* recta trahatur *BG* intercipiens

1 XCII: 91^a *B*
1–2 suma[tur . . . . punctum *illeg. D*
2–3 in . . . . ortho]gonium *illeg. D*
3 contentum: continentium *B*
4–5 C[irculo. . . .rect]a *illeg. D*
5 BG:GB *OP*
5–7 BG. . . .centrum *illeg. D*
7–9 D[A. . . .est[1] *illeg. D*
8–10 D. . . .punctorum *om. B*
9 est[1] *om. P*
9–10 circulus. . . .ZA *illeg. D*
10–11 Datum. . . .ZAE *om. B*
11 *de* Et est *add. mg. manu rec. O* per XXXIIII 3[1] euclidis / Datum: Datus *B*
11–12 est[1]. . . .BAG *illeg. D*
12 BAG: BA, BG *B*
13 XCIII: 92^a *B*
13–14 circulum . . . . portionem *illeg. D*
14–15 recipien[tem . . . . coutreque *illeg. D* / et . . . . datum *om. P*
15 coutreque: coutraque *B*
15–16 angulum. . . .habebunt *illeg. D*
17 sub *om. B* / continentium: continencium *BP*
17–18 et . . . . deorsum *illeg. D*
18 dividente: dividentem *B*
18–19 in[tercepta . . . . erit *illeg. D*
19 circumferentiam *correxi ex* circumferencium *OP* periferiam *B*
20–2(p.s.) circulo. . . .recta *illeg. D*

and ZD is given. Therefore $AD \times DZ$ is given. And $AD \times DZ = BD \times DG$. Therefore $BD \times DG$ is also given.

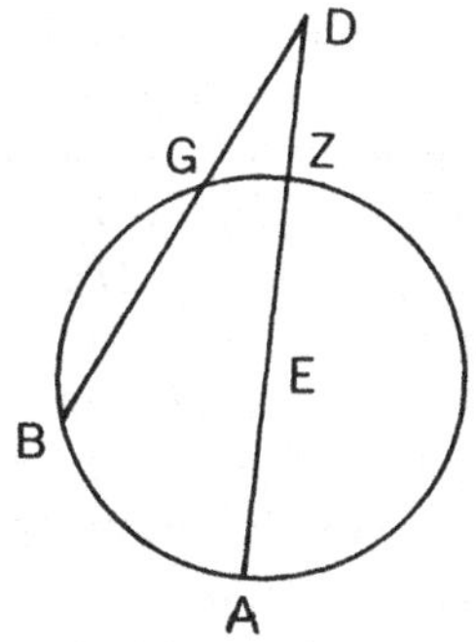

XCII. If in a circle given in position there is taken a given point and through that point there is drawn a straight line to the circle, the rectangle comprised by the segments of that line so drawn is given.

In a circle BG given in position, let there be taken a given point A and through that point A let there be drawn the straight line BG. I say that $BA \times AG$ is given.

Let D be taken as the center of the circle and let DA, when joined, be extended to the points E and Z. Since each of the points D and A is given, DA is given in position. But the circle GZB is also given in position. Therefore each of the points Z and E are given. But A is also given. Therefore each of ZA and AE is given. Therefore $ZA \times AE$ is given. And $ZA \times AE = BA \times AG$. Therefore $BA \times AG$ is also given.

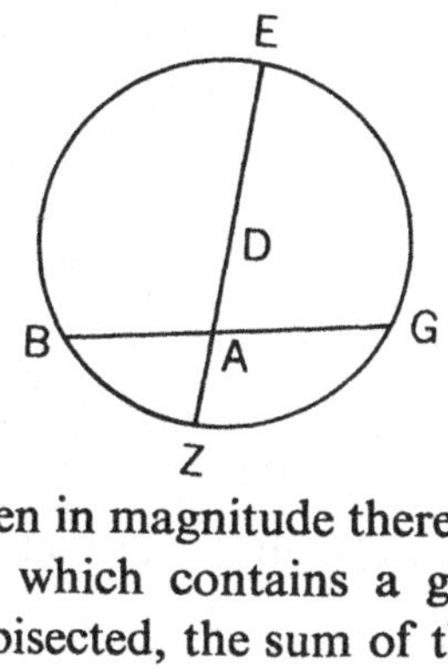

XCIII. If in a circle given in magnitude there is drawn a straight line intercepting a segment which contains a given angle, and if the angle in the segment is bisected, the sum of the straight lines includ-

portionem recipientem angulum datum *BAG* et dividatur *BAG* angulus in duo equa ab *AD* recta. Dico quod proportio est coutriusque *BAG* ad *AD* data et quoniam datum est quod sub coutraque *BAG* et *ED*.

Coniungantur *B*, *D*. Et quoniam in circulum datum magnitudine *DAG* protracta est recta *BG* intercipiens sectionem *BAG* recipientem angulum datum *BAG*, data ergo est *BG* magnitudine. Et per eadem quoque et *BD* data est magnitudine. Proportio ergo *BG* ad *BD* data. Et quoniam *BAG* angulus in quo equa dividitur ab recta *AD*, est ergo sicut *BA* recta ad *AG* ita *BE* ad *EG*. Permutatim ergo sicut *AB* ad *BE* ita *AG* ad *GE*. Et sicut ergo coutraque *BAG* ad *BG* ita *AG* ad *GE*. Et quoniam equalis est *BAE* angulus angulo *EAG*, est autem et *AGE* angulus angulo *BDE* equalis, reliquus ergo *AEG* reliquo *ABD* est equalis. Equiangulus ergo est *AEG* trigonus *ABD* trigono. Est ergo sicut *AG* ad *GE* ita *AD* ad *DB*. Sed sicut *AG* ad *GE* ita coutraque *BA*, *AG* ad *BG*. Est ergo sicut coutraque *BA*, *AG* ita *AD* ad *DB*. Permutatim ergo sicut coutraque *BAG* ad *AD* ita *BG* ad *BD*. Proportio vero *BG* ad *BD* est data. Proportio ergo et coutriusque *BAG* ad *AD* data.

Dico quoniam et quod sub coutraque *BAG* et *ED* datum est.

Quoniam enim equiangulus est *AEG* trigonus *DEB* trigono, est ergo sicut *BD* ad *DE* ita *AG* ad *GE*. Sicut vero *AG* ad *GE* ita est coutraque *BAG* ad *BG*. Et sicut ergo coutraque *BAG* ad *GB* ita est *BD* ad *DE*. Quod ergo sub coutraque *BAG* et *ED* equale est ei quod sub *GB*, *BD*. Datum vero quod sub *GB* et *BD*. Datum ergo et quod sub coutraque *BAG* et *ED*.

Aliter

Protrahatur *AG* super *E* iaceatque recte *BG* equalis *GE* et coniungantur

2–4 ad. . . . Et *illeg. D*
3 coutraque: utraque *P*
4–5 mag[nitudine. . . . angulum *illeg. D*
5 BAG:DAG *O*
6 ergo *tr. B post* est / BD: BAD *B*
6–7 data. . . . est *illeg. D*
7–8 ad. . . . dividi]tur *illeg. D*
8 angulus: angulum *P*
8–9 re[cta². . . . AB *illeg. D*
10 BAG. . . . BAE *illeg. D* / angulus *om. B*
11 *de* AGE *add. mg. manu rec. O* III, XXI
11–12 BDE. . . . Equi]angulus *illeg. D*
12 ergo² *tr. B post* est² / trigonus *om. B*
13 Est. . . . AG¹ *illeg. D*
14–15 si[cut . . . . Permutatim *illeg. D*
15–16 Proportio. . . . ergo *illeg. D*
16 coutriusque: coutraque *B*
18 quoniam *tr. P ante* quod
18–19 co[utraque. . . . est¹ *illeg. D*
20 ita. . . . ita *illeg. D*
21 GB: BG *B* / est *om. B*
21–22 ita. . . . coutraque *illeg. D*
22 et *om. D* / equale: equalis *D* / ei: e *B*
23 GB. . . . coutraque *illeg. D* / ED:DE *P*
25–1(p.s.) equa[lis. . . . est *illeg. D*

ing the given angle will have a given ratio to the line which bisects the angle; and the rectangle contained under both [1] the sum of the lines including the given angle and [2] the lower segment of the line bisecting the angle at the circumference will be given.

In the circle *ABG* given in magnitude, let the straight line *BG* be drawn intercepting the segment containing the given angle *BAG* and let the angle *BAG* be bisected by the straight line *AD*. I say that the ratio $(BA + AG)$: AD is given and that $(BA + AG) \times ED$ is given.

Let *B* and *D* be joined. Since in the circle *DAG* given in magnitude there is drawn the straight line *BG* intercepting the segment *BAG* which contains the given angle *BAG*, therefore *BG* is given in magnitude. By the same argument, *BD* is also given in magnitude. Therefore the ratio $BG{:}BD$ is given. Since the angle *BAG* is bisected by the straight line *AD*, hence $BA{:}AG = BE{:}EG$. Hence, by permutation, $AB{:}BE = AG{:}GE$. Therefore $(BA + AG){:}BG = AG{:}GE$. Since $\angle BAE = \angle EAG$ and $\angle AGE = \angle BDE$, the remaining angle *AEG* is equal to the remaining angle *ABD*. Therefore triangle *AEG* is equiangular with triangle *ABD*. Therefore $AG{:}GE = AD{:}DB$. Therefore $AG{:}GE = (BA + AG){:}BG$. Therefore $(BA + AG){:}BG = AD{:}DB$. Therefore, by permutation, $(BA + AG){:}AD = BG{:}BD$. But the ratio $BG{:}BD$ is given. Therefore the ratio $(BA + AG){:}AD$ is also given.

I say that $(BA + AG) \times ED$ is also given.

Since the triangle *AEG* is equiangular with the triangle *DEB*, hence $BD{:}DE = AG{:}GE$. But $AG{:}GE = (BA + AG){:}BG$. Therefore $(BA + AG){:}GB = BD{:}DE$. Hence $(BA + AG) \times ED = GB \times BD$. But $BG \times BD$ is given. Therefore $(BA + AG) \times ED$ is also given.

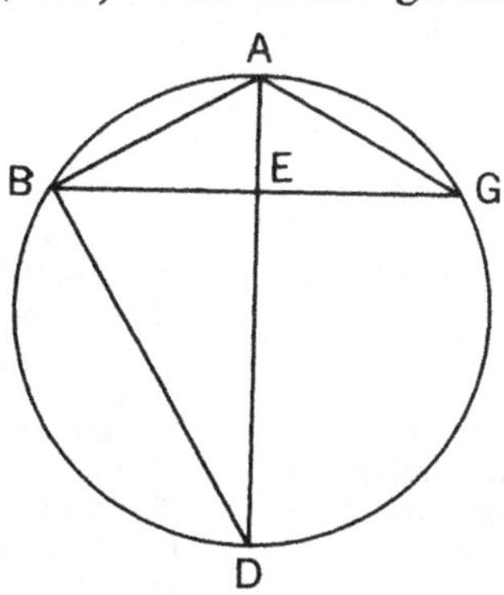

Another proof

Let *AG* be extended to *E* and let it be posited that $GE = BG$, and let *EB* and *BD* be joined.

EB, *BD*. Et quoniam duplus est qui sub *AGB* utriusque qui sub *AGD* et (72r) *GBE*, equalis est ergo *GBE* angulus angulo *AGD*, hoc est, *ABD*./Communis adiaceat *ABG* angulus. Totus ergo *DBG* toti *ZBE* est equalis. Est autem et *GAB* angulus angulo *GDB* equalis. Reliquus ergo *GEB* reliquo *DGB* est equalis. Equiangulus ergo est *EAB* trigonus *GDB* trigono. Est ergo sicut *EA* ad *AB* ita *GD* ad *DB*. Atqui *EA* coutraque est *AGB*. Sicut ergo coutraque *AGB* ad *AB* ita *GD* ad *DB*. Et permutatim ergo sicut coutraque *AGB* ad *GD* ita *AB* ad *DB*. Proportio vero *AB* ad *BD* data, utraque enim earum data. Proportio ergo et coutriusque *AGB* ad *GD* data. Et quoniam equiangulus est *EAB* trigonus trigono *ZBD*, est ergo sicut *EA* *AB* ita *BD* ad *DZ*. Atqui *EA* coutraque est *AGB*. Sicut ergo coutraque *AGB* ad *AB* ita *BD* ad *DZ*. Quod ergo a coutraque *AGB* et *ZD* equales est ei quod sub *ABD*. Datum vero est quod sub *ABD*, data enim utraque earum. Datum ergo et quod sub coutraque *AGB* et *ZD*.

Aliter

Protrahatur *AG* super *Z* iaceatque recte *AB* equalis *GZ* et coniungantur *BD*, *DG*, *DZ*. Quoniam ergo equalis est *BA* quidem recte *GZ*, atqui *DB* recta recte *DG*, due vero *AB*, *BD* duabus *ZG* et *GD* equales sunt utraque unicuique. Et angulus *ABD* angulo *DGZ* equalis est, quoniam quidem in circulo est *ABGD* quadrilaterum. Basis ergo *AD* basi *DZ* equalis et *ABD* trigonus *GDZ* trigono est equalis et reliqui anguli reliquis angulis, quibus equa latera subtenduntur. Equalis ergo *BDA* angulus *DZG* angulo. Datus vero *BAD* angulus. Datus ergo et *DZG* angulus. Est vero et *DAZ* angulus datus. Datus est ergo *ADZ* trigonus specie. Proportio ergo *ZA* ad *AD*

1 *post* quoniam *scr. B* ergo / *supra* duplus *add. O* per XXXII primi(?) / qui: quod *P*
2 GBE[2]....ABD *illeg. D*
3 DBG: BDG *B*
3–4 equalis....GDB *illeg. D*
4 GDB: GBD *B* / reliquo: relique *D*
5 est[2] *om. B* / Est *tr. B post* ergo *ln. 6*
5–6 ergo....sicut *illeg. D*
7 ergo[1]....Et *illeg. D*
7–8 AB....ad[1] *om. B*
8 *post* DB *scr. D* da[ta, utraque enim earum data(?) / Proportio....ad *illeg. D*
9–10 ad....EAB *illeg. D*
10 ZBD:ZDB *B*
11 est *om. B*
11–12 Atqui....A]GB[1] *illeg. D*
12–13 est....A]BD[2] *illeg. D*
13 est *om. B* / ABD[2]:AB *B*
15–16 Aliter....AB *illeg. D*
17 quidem: qua *B*
17–18 quidem....vero *illeg. D*
19 Et....est *illeg. D*
20–21 equa[lis....anguli *illeg. D*
22–23 subtendun[tur....BAD *illeg. D*
23–24 et[2]....Proportio *illeg. D*

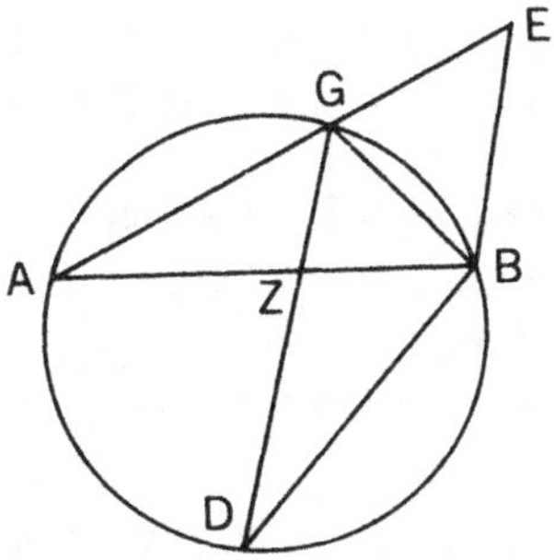

Since the angle AGB is double each of the angles AGD and GBE, hence $\angle GBE = \angle AGD$, that is, $\angle GBE = \angle ABD$. Let the common angle ABG be added. Therefore the whole DBG is equal to the whole ZBE. But $\angle GAB = \angle GDB$. Therefore the remaining angle GEB is equal to the remaining angle DGB. Therefore the triangle EAB is equiangular with the triangle GDB. Therefore $EA:AB = GD:DB$. Now $EA = AG + GB$. Therefore $(AG + GB):AB = GD:DB$. And, by permutation, $(AG + GB):GD = AB:DB$. But $AB:DB$ is given, for each of them is given. Therefore the ratio $(AG + GB):GD$ is given. Since the triangle EAB is equiangular with the triangle ZBD, hence $EA:AB = BD:DZ$. Now $EA = AG + GB$. Therefore $(AG + GB):AB = BD:DZ$. Therefore $(AG + GB) \times ZD = AB \times BD$. But $AB \times BD$ is given, for each of them is given. Therefore $(AG + GB) \times ZD$ is also given.

Another proof

Let AG be extended to Z and let it be posited that $GZ = AB$, and let BD, DG and DZ be joined.

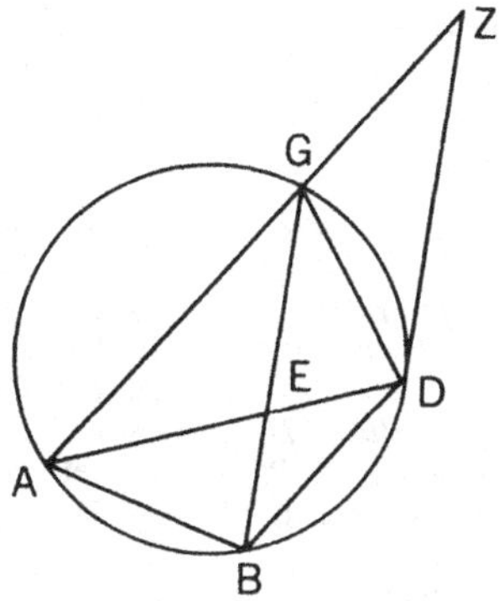

data. At vero *AZ* coutraque est *BAG*, quoniam equalis est *GZ* recte *BA*. Proportio ergo coutriusque *BAG* ad *AD* data. Et similiter primo demonstrabimus quoniam quod sub coutraque *BAG* et *ED* datum est.

XCIIII. Si circulo dato positione super diametrum datus punctus sumatur, ab eo vero puncto ad circulum assumatur aliqua recta, et a sectione perpendicularis trahatur illi protracte, per punctum vero in quo coniungitur perpendicularis periferie parallela trahatur ei protracte, datus est punctus in quo coniungitur parallela illa diametro et quod sub parallelis orthogonium continetur datum erit.

Circulo enim positione dato *ABG* super diametrum *BG* sumatur datus punctus *D* et per eum ad circulum producatur quedam fortuitu *DA*, a puncto vero *A* recte *DA* ad rectos angulos recta trahatur *AE*, per *E* vero recte *AD* parallela trahatur *EZ*. Dico quod datus est punctus *Z* et quoniam quod sub *AD* et *EZ* spatium datus est.

Protrahatur *EZ* super *T* et coniungatur *A*, *T*. Et quoniam rectus est *TEA* angulus, *TA* diametros est *ABG* circuli. Est vero et *BG*. Ergo *I* centrum est *ABG* circuli. Datus est ergo punctus *I*. Est autem et *D* datus. Data est ergo *DI* magnitudine. Et quoniam parallela est *AD* ei que est *ET* et est equalis *TI* ei que est *IA*, equalis ergo et *DI* quidem recte *IZ*, et vero *AD* ei que est *ZT*. Data vero *DI*. Data ergo *ZI*. Sed et positione. Utraque ergo *IZ* et *ID* data est. Et est datus *I*. Datus ergo et *Z*. Et quoniam circulo

1–2 BAG. . . .coutriusque *illeg. D*
2–3 demonstrabi[mus. . . .est *illeg. D*
4 XCIIII *om. B* / Si . . . . p[unctus *illeg. D*
5 eo . . . . recta *illeg. D*
6 a *om. B* / perpen[dicularis. . . .vero *illeg. D*
6–8 per . . . . protracte *om. B*
7–8 perpendicularis . . . . da]tus *illeg. D*
7–9 in . . . . ortho]gonium *illeg. D*
9 parallelis *correxi ex* paralellis *O* parallilis *P* paralillis *B* / continetur *om. B*
9–10 con[tinetur . . . . su]matur *illeg. D*
10 datus tr. *B post* punctus *ln. 11*
10–11 datus. . . .queda]m *illeg. D*
11–4(p.s.) *pro* producatur. . . .magnitudinibus *substituit B verba alterius operis*
11–12 for[tuitu. . . .rec]te *illeg. D*
13–14 para[llela. . . .datus *illeg. D*
15–16 EZ. . . .vero *illeg. D*
16–18 Ergo. . . .magni]tudine *illeg. D*
18–19 Et. . . .ve]ro *illeg. D*
19 et[2] *correxi ex* at *OP*
20–21 ei. . . .ergo[2] *illeg. D*
21–2(p.s) Et[2]. . . .ergo[1] *illeg. D*

Since $BA = GZ$ and $DB = DG$, the two straight lines AB and BD are equal to ZG and GD, each to its corresponding line. And $\angle ABD = \angle DGZ$, since the quadrilateral $ABGD$ is inscribed in the circle. Therefore the base AD is equal to the base DZ, the triangle ABD is equal to the triangle GDZ and the remaining angles are equal to the remaining angles, *i.e.* those by which the equal sides are subtended. Therefore $\angle BDA = \angle DZG$. Now $\angle BAD$ is given. Therefore $\angle DZG$ is also given. But $\angle DAZ$ is also given. Therefore the triangle ADZ is given in species. Therefore the ratio $ZA{:}AD$ is given. But $AZ = BA + AG$, since $GZ = BA$. Therefore the ratio $(BA + AG){:}AD$ is given. As in the first proof, we shall demonstrate that $(BA + AG) \times ED$ is given.

XCIV. If on the diameter of a circle given in position there is taken a given point, and from that point there is drawn some straight line to the circle, and if from the point of intersection of the line there is drawn a perpendicular to the line so drawn, and if through the point at which that perpendicular meets the circumference there is drawn a parallel to the line so drawn, then the point at which that parallel meets the diameter is given and the rectangle contained by the parallel lines will be given.

On the diameter BG of the circle ABG given in position let there be taken the given point D and through it to the circle let there be drawn any straight line DA, and from the point A there be drawn a straight line AE at right angles to the straight line DA, and through E let there be drawn the line EZ parallel to the straight line AD. I say that the point Z is given and that the space $AD \times EZ$ is given.

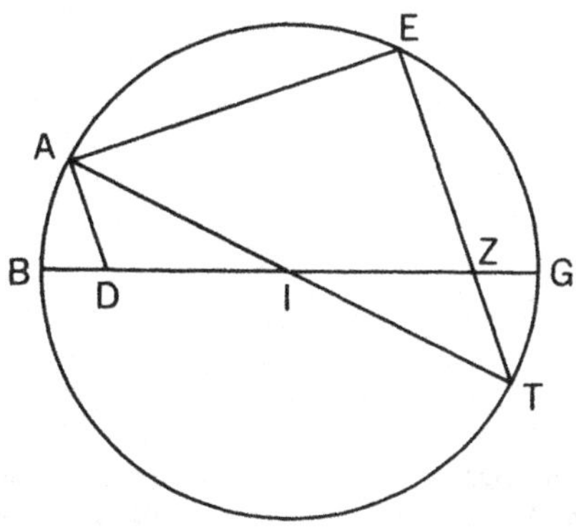

dato positione *ABG* sumptus est punctus datus *Z* et protracta est *EZT*, datum ergo quod sub *EZT*. Equalis vero *TZ* ei que est *DA*. Datum ergo quod sub *AD* et *EZ*.

Explicit liber de datis magnitudinibus.

2 DA *correxi ex* TA *DOP*
3 sub....EZ *illeg. D*
4 liber....magnitudinibus *om. P* / datis magnitudinibus *illeg. D*

Let the straight line EZ be extended to T and let A and T be joined. Since $<TEA$ is a right angle, TA is the diameter of the circle ABG. But BG also is the diameter. Therefore I is the denter of the circle ABG. Therefore the point I is given. Now D is also given. Therefore DI is also given in magnitude. Since AD is parallel to ET and $TI = IA$, therefore $DI = IZ$ and $AD = ZT$. But DI is given. Therefore ZI is given. But it is also given in position. Therefore each of IZ and ID is given. And I is given. Therefore Z is also given. Since in the circle ABG given in position the given point Z has been assumed and EZT is drawn, therefore $EZ \times ZT$ is given. But $TZ = DA$. Therefore $AD \times EZ$ is given.

Here ends the book on given magnitudes.

CHAPTER 6

NOTES ON THE TEXT

1) Def. XI & XII, p. 56, ln. 7–10

These definitions are a little difficult to understand, because the meaning of the peculiar phrase "as in ratio" (*quam in proportione*) is not clear at a glance. As a matter of fact, Simson omitted these definitions, because he thought that "they could not be given in English so as to have any tolerable sense." Cf. R. Simson, *The Elements of Euclid*, p. 464. In view of the later actual use of this notion after Proposition X, however, we can understand that "to be greater or less *as in ratio*" means the following:

A magnitude A is greater than another magnitude B *as in ratio*, when $(A - \delta): B = k_1; k_2$ (δ = a given magnitude, $k_1: k_2$ = a given proportion), while A is less than B *as in ratio*, when $(A + \delta): B = k_1: k_2$.

Therefore, I have added in translation the phrase "the excess of A over a given quantity has a given proportion to B" to explicate the literal translation "A is greater than B by a given quantity as in ratio."

The above interpretation is also suggested by an ancient scholium on this definition. Cf. *Euclidis Opera Omnia*, Vol. VI, p. 263. *Scholium antiquum ad def. XI*, "*παρὰ τοσοῦτον γὰρ οὐκ ἔχουσι λόγον δοθέντι τὰ δύο μεγέθη, παρ' ὅσον ὑπερέχει τὸ ἓν τοῦ ἑτέρου δοθέντι τινὶ μεγέθει, οὗ ἀφαιρεθέντος εὑρίσκεται καὶ ὁ δεδομένος λόγος τῶν δύο μεγεθῶν.*

2) Def. XIII, XIIII & XV, p. 56, ln. 11–15

These three definitions concerning the Greek terms *catigmeni, anigmeni*, and *parathesi* derived from Apollonius, as is mentioned in an ancient scholium. Cf. *Euclidis Opera Omnia*, Vol. VI, p. 264. *Scholium ad def. XIII*, "*τούτους Ἀπολλωνίου φασὶν εἶναι τοὺς τρεῖς ὅρους.*"

The meaning of these notions can be understood as follows:

Actually, there are three kinds of ways in which a straight line is drawn from a given point to a straight line given in position.

The first case is that a line AB is drawn *down* from a given point A to a given line L like 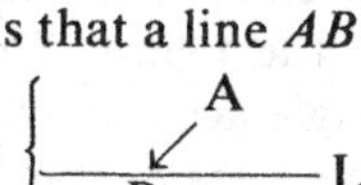

(catigmenni = *κατα-ἡγμένη*).

The second case is that a line AB is drawn *up* from a given point A to

a given line *L* like 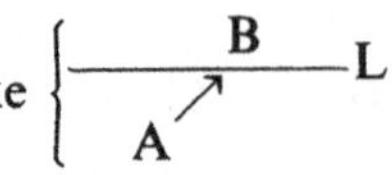

(anigmeni = ἀνα-ἡγμένη).

The third is that a line *AB* is drawn from a given point *A parallel* to a given line *L* like 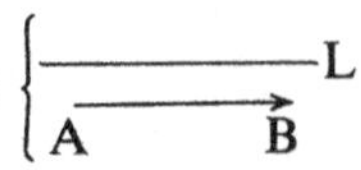

(parathesi = παρα-θέσει).

Simson omitted these definitions XIII, XIV, and XV, because he thought that they were of no use.

By the way, Menge's translation is no good here, for he translated ἐπί as *ad* in Definition 13 and πρός as *in* in Definition 14, while: they should be *super* and *ad* respectively.

3) Prop. XXVIII, p. 98, ln. 1 & 3

The Greek rendered by *secundum positionem* is παρὰ θέσει which means "parallel." This Latin translation, therefore, is too literal. So I translated this not as "following position" but as "parallel."

4) Prop. LII, p. 142, ln. 1

Tetragonus (τετράγωνον) here means "square" and not "four-sided figure" in general. This special usage began with Pythagoreans (cf. Aristotle, *Metaphysics*, 986a 26), and it is most commonly used by Aristotle. But Greek mathematicians and philosophers sometimes used this term to mean any quadrilateral. Euclid avoided this ambiguity by introducing another word τετράπλευρον for a quadriliteral. See Definition XIX of the First Book of the *Elements*. Cf. Heath, *The Thirteen Books of Euclid's Elements*, Vol. I, pp. 187–188.

5) Prop. LX, p. 154

The word *gnomon* (γνώμων), which means literally a thing enabling something to be known, was first used in the sense in which Herodotus told us (II. 109) that "the Greek learnt the πόλος, the *gnomon*, and the twelve parts of day from the Babylonians." This *gnomon* refers to a kind of sundial which measured time by shadows produced by the sun.

Secondly, we find that this term was used for a mechanical instrument for drawing right angles, as is shown in the figure. This *gnomon* was a carpenter's square for drawing a right angle accurately.

Thirdly, the *gnomon*, owing to its shape, became the geometrical figure which remained of a square when a smaller square was taken out of one corner, or, "the figure, as Aristotle says, which when added to a square

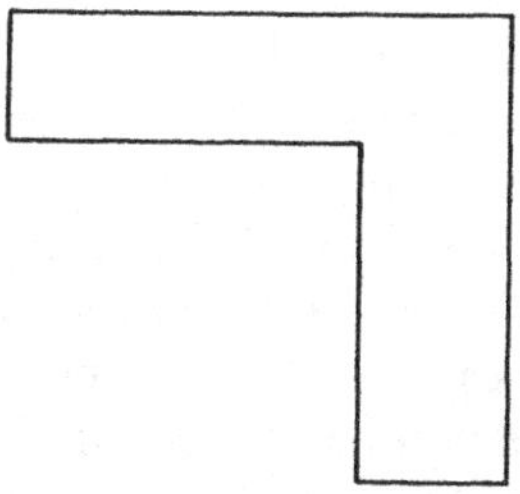

increases its size but does not alter its form." As Heath says, "the Pythagoreans used the term in this sense, and further applied it by analogy to the series of odd numbers as having the same property in relation to square numbers."

Fourthly, Euclid extended the notion of *gnomon* so that it had the same relation to any *parallelogram* as it had to a square. When he says that "if a parallelogram is augmented or diminished by a given *gnomon*. . . .", he used, of course, this word in his own fourth meaning. See Definition II of the Second Book of the *Elements*.

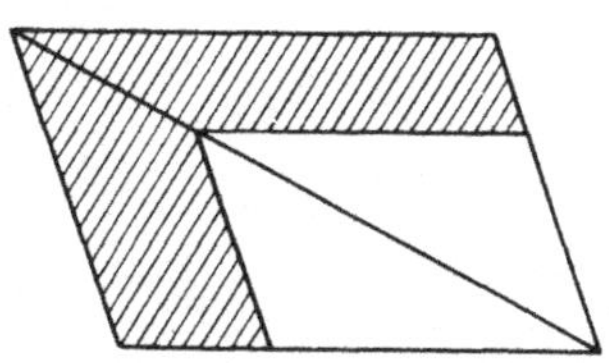

Later, Hero of Alexandria gave a further extended definition of a *gnomon in general* as "any figure which, when added to any figure whatever, makes the whole figure similar to that to which it is added."

Cf. Heath, *Op. cit.*, pp. 370–371.

6) Prop. LXXXIV & LXXXV, pp. 204–207

These propositions can be interpreted as the solution of the simultaneous equations:

$$\left.\begin{aligned} x \pm y &= a \\ xy &= b \end{aligned}\right\}$$

This is equivalent to the solution of the mixed quadratic equations $x^2 \pm bx = c$ in Propositions IV and V of the Second Book of the *Elements* and shows another way of geometrical solution of quadratic equations by the method of "application of areas."

For details, cf. Health, *Op. cit.*, pp. 379–385

7) Prop. XCI & XCII, pp. 214–217

These propositions are equivalent to Propositions XXXV and XXXVI of the Third Book of the *Elements* about the "power" of a point with reference to a circle to the effect that "given a circle and a point in the same plane, the rectangle contained by the intercepts between this point and the points in which respectively the circumference is cut by any straight line passing through the point and meeting the circle is also given."

Cf. Heath, *A History of Greek Mathematics*, Vol. I, p. 423

BIBLIOGRAPHY

A. Manuscripts

Oxford, Bodleian Library, Auctarium, F.5.28, 13c, 99r-113r. (Data)
Paris, Bibliothèque Nationale, Fonds latin 16648, 13-14c, 60r-91v. (Data)
Dresden, Sächsische Landesbibliothek, Db 86, 14c, 200r-213r. (Data)
Berlin, Staatsbibliothek, Preußischer Kulturbesitz, Q510, 14c, 178v-191r. (Data)
Oxford, Bodleian Library, Auctarium, F.5.28, 13c, 57r–64r. (Optics)
Oxford, Bodleian Library, Auctarium, F.5.28, 13c, 64r–69r. (Catoptrics)
Paris, Bibliothèque Nationale, Fonds latin 7373, 13c, 2r-167v. (Elements)

B. Books and Periodicals

Barrow, Isaac. *The Elements of Euclid. The whole Fifteen Books compendiously demonstrated with Archimedes' Theorem of the Sphere and Cylinder investigated by the Method of Indivisibles by Isaac Barrow D. D. To which is annexed Euclid's Data and a brief Treatise of Regular Solids.* London, 1732.

Björnbo, Axel A. "Die mettelalterlichen Übersetzungen aus dem Griechischen auf dem Gebiete der mathematischen Wissenschaften", *Archiv für die Geschichte der Naturwissenschaft und der Technik*, Vol. I (1909), pp. 385–394.

Boese, Helmut. *Die mittelalterliche Übersetzung der ΣΤΟΙΧΕΙΩΣΙΣ ΦΥΣΙΚΗ des Proclus.* Berlin, 1958.

Bubnov, Nikolai. *Gerberti postea Silvestri II papae Opera Mathematica.* Berlin, 1899.

Buchbinder, Friedrich. *Euklids Porismen und Daten.* Nürnberg, 1825.

Busard, H. L. L. *The Translation of the Elements of Euclid from the Arabic into Latin by Hermann of Carinthia(?).* Leiden, 1968.

_________, *The Translation of the Elements of Euclid from the Arabic into Latin by Hermann of Carinthia(?) Books VII-XII.* Amsterdam, 1977.

Cantor, Moritz B. *Vorlesungen über Geschichte der Mathematik.* 4 vols. Leipzig, 1894–1908.

Cappelli, Adriano. *Dizionario di Abbreviature latine ed italiane.* Milano, 1961.

Carmody, Francis J. *The Astronomical Works of Thabit B. Qurra.* Berkeley, 1960.

_________. *Arabic Astronomical and Astrological Sciences in Latin Translation.* Berkeley, 1956.

Clagett, Marshall. *The Science of Mechanics in the Middle Ages.* Madison, 1959.

_________. *Archimedes in the Middle Ages.* Vol. I, Madison, 1964; Vol. II-III, Philadelphia, 1976–1978.

_________. "The Medieval Latin Translation from the Arabic of the Elements of Euclid, with Special Emphasis on the Versions of Adelard of Bath", *Isis*, Vol. XLIV (1953), pp. 1642.

_________. "Medieval Mathematics and Physics: A Check List of Microfilm Reproductions", *Isis*, Vol. XLIV (1953), pp. 371–381.

_________. "King Alfred and the Elements", *Isis*, Vol. XLV (1954), pp. 269–277.

_________. "Medieval Mathematics, Physics and Philosophy: A Revised Catalogue of Photographic Reproductions", *Manuscripta*, II (1958), pp. 131–154 and III (1959), pp. 19–38.

_________. *Euclid's Elements in the Middle Ages* (Unpublished address given at Dumbarton Oakes College). Washington, 1959.

_________. "Some General Aspects of Medieval Physics", *Isis*, Vol. XXXIX (1948), pp. 29–36.

Delisle, Leopold V. *Le Cabinet des Manuscrits de la Bibliothèque Nationale*. 3 vols. and 1 suppl. Paris, 1868–81.

Euclid. *Euclidis Opera Omnia*. 8 vols. and 1 suppl. eds. J. L. Heiberg and H. Menge. Leipzig, 1883–1916.

_________. *Euclidis Elementa*, Vol. I—III, ed. E. S. Stamatis. Leipzig, 1969–1972

Folkerts, Menso. *"Boethius" Geometrie II: Ein mathematisches Lehrbuch des Mittelalters*, Wiesbaden, 1970.

_________. *Ein neuer Text des Euclids Latinus: Faksimiledruck der Handschrift Lüneburg D 4° 48, f. 13r-17v*, Hildesheim, 1970.

Goldat, George D. *The Early Medieval Tradition of Euclid's Elements*. Madison, 1956. (Unpublished dissertation)

Gow, James. *A Short History of Greek Mathematics*. A reprint, New York, 1923.

Gregory, David. *Ευκλείδου τὰ σωζόμενα, Euclidis quae supersunt omnia. Ex recensione Davidis Gregorii M. D. Astronomiae Professoris Saviliani et R.S. S. Oxoniae, e Theatro Sheldoniano*. Oxford, 1703.

Hankel, Hermann. *Zur Geschichte der Mathematik in Altertum und Mittelalter*. Leipzig, 1874.

Hartwig, O. "Die Übersetzungsliteratur Unteritaliens in der normannisch-staufischen Epoche", *Centralblatt für Bibliothekswesen*, Vol. III (1886), pp. 161–190, 223–225, 505 seq.

Haskins, Charles H. *Studies in the History of Medieval Science*. 2nd ed. Cambridge, Mass., 1921.

Haskins, Charles H. *The Renaissance of the Twelfth Century*. Cleveland-New York: Meridian Books, 1957.

_________. *The Normans in the European History*. Boston, 1915.

_________ and Lockwood, P. "The Sicilian Translators of the Twelfth Century and the First Version of Ptolemy's *Almagest*", *Harvard Studies in Classical Philology*. Vol. XXI (1910), pp. 75–102.

_________. "Further Notes on Sicilian Translators of the Twelfth Century", *Ibid.*, Vol. XXIII (1912), pp. 155–166.

Heath, Sir Thomas L. *The Thirteen Books of Euclid's Elements.* 3 vols. New York: Dover edition, 1956.

_________. *A History of Greek Mathematics.* 2 vols. Oxford, 1927.

_________. *A Manual of Greek Mathematics.* New York: Dover edition, 1963.

Heiberg, Johan L. *Litterargeschichliche Studien über Euklid.* Leipzig, 1882.

_________. "Philologische Studien zu griechischen Mathematikern III. Die Handschriften Georg Vallas von griechischen Mathematikern", *Jahrbücher für classische Philologie*, Suppl. Vol. XII (1881), pp. 377–402.

_________. "Eine mittelalterliche Übersetzung der Syntaxis des Ptolemaios", *Hermes*, Vol. XLV (1910), pp. 57–66.

_________. "Noch einmal die mittelalterliche Ptolemaios-Übersetzung", *Ibid.*, Vol. XLVI (1911), pp. 207–216.

_________. "Beiträge zur Geschichte der Mathematik im Mittelalter", *Zeitschrift für Mathematik und Physik*, Vol. XXXV (1890), historisch-literarische Abteilung, pp. 45–58, 81–98.

_________. "Les sciences grecques et leur transmission", *Scientia* Vol. XXXI (1922), pp. 1–10, 97–104.

Hultsch, Friedrich, "Eukleides", Pauly-Wissowa's *Die Real-Encyclopädie der classischen Altertumswissenschaften.* New edition, Vol. VI, Part 1, Stuttgart (1907), pp. 1003–1052.

Ito, Shuntaro, *The Medieval Latin Translation of the Data of Euclid.* Madison, 1964. (Unpublished dissertation)

_________. "On the Medieval Latin Translation of the *Data* of Euclid", *Japanese Studies in the History of Science*, No. 5 (1966). pp. 100–118.

_________. "On a Manuscript of the *Elements* of Euclid in the 14th Century—MS Paris, Bibl. Nat. Fonds latin 7215" (in Japanese with English summary) *The Proceedings of the Department of Humanities, College of General Education, University of Tokyo*, Vol. XIII (1965). pp. 1–32.

_________. "Euclid—His Works and Their Tradition" (in Japanese), *The Journal of the Department of Liberal Arts, University of Tokyo*, No. 3 (1970). pp. 13–56.

_________. "The Origin of Pure Mathematics—with Reference to Formation of Euclid's *Elements*" (in Japanese), *Shiso*, No. 513 (1967). pp. 39–56.

_________. *Science in Civilizations* (in Japanese), Tokyo. 1976.

_________. *Roots of Modern Science* (in Japanese), Tokyo, 1978.

Ito, S., Hara, K. and Murata, T. *A History of Mathematics* (in Japanese), Tokyo, 1975.

Iyanaga, S., Ito, S. and Sato, T. *Greek Mathematics* (in Japanese), Tokyo, 1979.

Lejeune, Albert. *L'optique de Claude Ptolémée.* Louvain, 1956.

Madan, Falconer C. et al. *A Summary Catalogue of Western Manuscripts in the Bodleian Library at Oxford.* 7 vols. in 8 tomes. Oxford, 1922–53.

Minio-Paluello, Lorenzo. "Henri Aristippe, Guillaume de Moerbeke et les traductions latines médiévales des « Météorologique » et du « De generatione et corruptione » d'Aristote", *Revue philosophique de Louvain*, Vol. XLV (1947), pp. 206–235.

________. "Les 'trois rédaction' de la traduction médiévale gréco-latine du « De generatione et corruptione » d'Aristote", *Ibid.*, Vol. XLVIII (1950), pp. 249–259.

________. "Gulielmo di Moerbeke, Traduttore della Poetica di Aristotele", *Rivista di Filosofia neo-scolastica*, Anno XXXIX—Fasc. I (1947), pp. 3–19.

________. "Note sull'Aristotele latino medievale, I-III", *Ibid.*, Anno XLII—Fasc. III (1950), pp. 1–16.

________. "Note sull'Aristotele latino medievale, V-VII", *Ibid.*, Anno XLIV (1952), pp. 389–495.

________. "Note sull'Aristotele latino medievale, VIII-IX", *Ibid.*, Anno XLVI—Fasc. III (1954), pp. 211–231.

________. "Le texte du « De anima » d'Aristote, La traduction latine avant 1500", *Autour d'Aristote* offert à Monseigneur A. Mansion. Louvain 1955, pp. 217–243.

Moody, Ernest A. and Clagett, Marshall. *The Medieval Science of Weights (Scientia de ponderibus)*. Madison, 1952.

Morrow, Glenn L. *Proclus A Commentary on the First Book of Euclid's Elements*, Princeton, 1970.

Murdoch, John E. "Euclides Graeco-Latinus", *Harvard Studies in Classical Philology*, Vol. 71 (1967), pp. 249–302.

________. "The Medieval Euclid: Salient Aspects of the Translations of the *Elements* by Adelard of Bath and Campanus of Novara", *Revue de Sythèse*, Tome LXXXIX (1968), pp. 67–94.

________. "Euclid: Transmission of the *Elements*", *Dictionary of Scientific Biography*, IV, New York, 1971, pp. 437–459.

Nakamura, K., Terasaka, H., Ito. S. and Ikeda, M. *Euclid Genron* (Japanese translation of the *Elements* of Euclid from the Greek), Tokyo, 1961.

Pappus, *Pappi Alexandrini collectio quae supersunt.* 3 vols. ed. Fr. Hultsch. Berlin, 1876–78.

Peyrard, F. *Euclidis quae supersunt. Les Oeuvres d'Euclide, en Grec, en Latin et en Français d'après un manuscrit très anciens, qui était resté inconnu jusqu'à nos jours.* 3 vols. Paris, 1814–1818.

Proclus. *Procli Diadochi in primum Euclidis elementorum librum commentarii.* ed. G. Friedlein. Leipzig, 1878.

Rose Valentin. "Die Lücke im Diogenes Laertius und der Alte Übersetzer", *Hermes*, Vol. I (1866), pp. 367–397.

Sarton, George. *Introduction to the History of Science.* 3 vols. in 5 tomes. Baltimore, 1927–1948.

Schwab, Johann Christoph. *Euklids Data, verbessert und vermehrt von Robert Simson*. Stuttgart, 1780.

Simson, Robert. *The Elements of Euclid, viz. the first six Books together with the eleventh and twelfth. In this Edition the Errors, by which Theon or others have long ago vitiated these Books, are corrected, and some of Euclid's Demonstrations are restored. Also The Book of Euclid's Data in the same manner corrected by Robert Simson*. 8th ed., Edinburgh, 1791.

Steinschneider, Moritz. *Die arabischen Übersetzungen aus Griechischen*. New edition. Graz, 1960.

_________. *Die europäischen Übersetzungen aus Arabischen bis zum 17. Jahrhundert*. New edition. Graz, 1956.

Suter, Heinrich. "Die Mathematiker und Astronomer der Araber und ihre Werke", *Abhandlungen zur Geschichte der Mathematik*, Vol. X (1900), pp. 1–278.

_________. "Das Mathematiker-Verzeichnis im Fihrist des Ibn Ja'kub an-Nadim", *Ibid.*, Vol. VI (1892), pp. 1–87.

Tannery, Paul. *La géométrie grecque: comment son histoire nous est parvenue et ce que nous en savons*. Paris, 1887.

_________. *Mémoires scientifiques*. 17 vols. Toulouse and Paris, 1912–1950.

Thaer, Clemens. *Die Data von Euklid*, Berlin-Göttingen-Heidelberg, 1962.

_________. "Euklids Data in Arabischer Fassung", *Hermes*, Vol. 77 (1942), pp. 197–205.

Theisen, W. R., *The Medieval Tradition of Euclid's Optics*. Madison, 1972. (Unpublished Dissertation)

Thorndike, Lynn. *A History of Magic and Experimantal Science*. 8 vols. New York, 1923–1958.

Valla, Georgius, *De expetendis et fugiendis rebus opus, Venice*, 1501.

Van Ryzin, Sister Mary St. Martin. *The Arabic-Latin Tradition of Euclid's Elements in the Twelfth Century*. Madison, 1960. (Unpublished dissertation).

Ver Eecke, Paul. *Pappus d'Alexandrie, La collection mathématique*. 2 vols. Paris-Bruges, 1933.

_________. *Proclus de Lycée, Les commentaires sur le premier livre des éléments d'Euclide*. Paris-Bruges, 1948.

_________. *Euclide, L'optique et la catoptrique*. Paris, 1959.

_________. *Les oeuvres complètes d'Archimède suivies des commentaires d'Eutocius d'Ascalon*. 2 vols. Paris, 1960.

Weissenborn, H. *Die Übersetzungen des Euklid durch Campano und Zamberti*. Halle, 1882.

Zamberti, Bartholomeus. *Elementa Euclidis megarensis philosophi platonicj mathematicarum disciplinarum Janitoris*, Venice, 1505.

Index of Names and Works

Index of Latin Terms

This index includes all Latin words found in the text. The references are to page (in bold type) and line numbers. Superscript numerals indicate the number of times the word occurs in a line. For each Latin word the corresponding Greek term in Menge's Greek text is indicated. The expression "no Gr equivalent" is used where there is no corresponding Greek word. Participles and other variations are reduced to the first-person singular verb form.

Made in the USA
Monee, IL
13 September 2026

62976342R00149